Reprints of Economic Classics

SCIENCE

The False Messiah

&

HOLIER THAN THOU

CLARENCE E. AYRES

SCIENCE
THE FALSE MESSIAH
[1927]

&

HOLIER THAN THOU
THE WAY OF THE RIGHTEOUS
[1929]

IN ONE VOLUME

WITH A NEW INTRODUCTION

Prolegomenon to Institutionalism

AUGUSTUS M. KELLEY • PUBLISHERS
CLIFTON 1973

Science: The False Messiah

First Edition 1927

(*Indianapolis*: The Bobbs-Merrill Company Publishers, 1927)

Holier Than Thou

First Edition 1929

(*Indianapolis*: The Bobbs-Merrill Company Publishers, 1929)

Reprinted *1973* by

AUGUSTUS M. KELLEY PUBLISHERS

Clifton   New Jersey   07012

Library of Congress Cataloging in Publication Data

Ayres, Clarence Edwin, 1891-
  Science: the false messiah & Holier than thou.

  (Reprints of economic classics)
  Reprints of the 1927 and 1929 editions, respectively, with new introd.
  1. Science—Philosophy. 2. Inventions.
3. Religion and science—1926-1945. 4. Ethics.
5. Social ethics. I. Ayres, Clarence Edwin, 1891- Holier than thou. 1973. II. Title. III. Title: Holier than thou.
B67.A79   215   71-130660
ISBN 0-678-00774-8

PRINTED IN THE UNITED STATES OF AMERICA
by SENTRY PRESS, NEW YORK, N. Y. 10013

PUBLISHER'S NOTE

The publisher wishes to point out that the introduction to the present reprint of the extremely rare *Science the False Messiah* [1927] and *Holier than Thou* [1929] is the last writing prepared for publication by the late Professor Clarence E. Ayres. I wish to thank Professor Joseph Dorfman for securing the introduction. He was the first economist to call attention to the significance of Professor Ayres's work, in volume IV of his *The Economic Mind in American Civilization*. [1050].

PROLEGOMENON TO INSTITUTIONALISM

When I was writing these two books, I had no thought of contributing to the literature of Institutionalism. Walton H. Hamilton had already given this conception of the economy its name, and I had been his admiring assistant (in current jargon, his "T.A.") when he did so. But I had taken a degree at Chicago in philosophy, had taught in the philosophy departments of Amherst and Reed, and still thought of myself (as perhaps I should still do) more as a would-be philosopher than as even a would-be economist. Nevertheless, as I look back over the lapse of nearly half a century, an unmistakable institutionalist pattern seems to stand out in bold relief.

Institutionalism has always seemed to me to derive its distinctive pattern from Thorstein Veblen, to whose earlier works Hamilton had introduced me when I was first associated with him in 1915. Others shared Veblen's tough-minded realism and his impatience with the phantasies of price theory. But insofar as Institutionalism is characterized by a distinctive conception of the industrial economy—and of human behavior and organized society generally—it is Veblen's conception.

These two conceptions—of "sportsmanship" as he called it in *The Theory of the Leisure Class* (1899) and "workmanship" as he called it everywhere—are woven like colored threads into all the products of his loom. Readers who are familiar only with *The Theory of the Leisure Class* have often failed to realize that "the instinct of workmanship" is present even in that book as a necessary foil to the "leisure class" proclivities with which the book is chiefly concerned. That is why Veblen's admirers continue to cite Lester F. Ward's long and admiring review of *Leisure Class*, written in 1900, soon after the publication of this classic-to-be. Not only did Ward recognize the significance of Veblen's references to "the instinct of workmanship"; he suggested that Veblen might well write an equally significant book on that theme. As everybody knows, Veblen did so—in 1914.

These are not textbooks. Veblen did not present students with neat definitions or tabulations of main points and subordinate points. He never said whether or not "the parental bent" and "idle curiosity" are manifestations of "the instinct of workmanship"—or *vice versa*. He never even attempted a precise definition of what seems to the casual reader to have been his central or basic concept.

But these deficiencies have not prevented the gradual development of a consensus. It is now generally understood that human behavior—all human behavior and all organized social patterns—present two distinct and contrasting (though not unrelated) aspects to the uninvolved observer. From the earliest times of which we have any knowledge down to the present human behavior presents an amazing contrast of rationality and irrationality, of sense and nonsense, of economy and waste. This contrast was Veblen's master principle. As the contrast between "making things" and "making money" it was the master principle of his economics and perhaps his most significant contribution to the dismal science.

That he was sometimes carried away by it seems all too evident to a later generation. His weakness for the "dolicho-blond race" strongly suggests his own filial piety as a Norwegian, and his idealization of "the savage state," in contrast to barbarian brutality clearly seems to reflect the utopianism of Edward Bellamy's *Looking*

Backward (by which in his youth, according to Joseph Dorfman, in *Thorstein Veblen and His America,* he had been fascinated) as well as Rousseau's "noble savage." Nevertheless, the fact remains that ancient mythologies persist in juxtaposition to the most advanced technology, such as a manned spacecraft hurtling back to earth from a landing on the moon.

Veblen made the dichotomy of technology and ceremonialism his master principle; and in my later efforts to clarify the "Institutionalist conception" of economic process and economic policy I have followed his lead. Was I doing so in the late twenties when these two books were written? Not consciously, I think. Quite clearly I was, in the mid-thirties when I wrote the article entitled "The Gospel of Technology" for publication in *American Philosophy Today and Tomorrow,* edited by Horace Kallen and Sidney Hook. But I do not think that is true of these volumes.

On this account I am mildly astonished to find that *Science the False Messiah* is concerned with one aspect of human nature and of all organized society, and *Holier Than Thou* with the other. Neither is so identified, and this I do not understand. For I was already an avowed Veblenian, and would not have hesitated to identify my ideas with his. But perhaps the explanation of this anomaly is simply that at this time my ideas were too inchoate to be identified even in my own mind with the dichotomy which later seemed so clear. Are they too confused to be sorted out? Perhaps it no longer matters. But for what it is worth, let me try.

SCIENCE THE FALSE MESSIAH

Although we no longer think in terms of instincts (such as the "instinct of workmanship") Veblen was certainly right in rooting the propensity to use tools, and therefore to make things, in the species endowment of mankind. No doubt the most important feature of this endowment is brain power. But scarcely less important is erect posture, in consequence of which the fore-paws, with their uniquely rich enervation and finger flexibility as well as the opposable thumb (which we share with many other species), all of which adds up to the inveterate fingering and handling which is so characteristic of our species. As the saying goes, "Satan finds work for idle hands to do."

Furthermore, in an environment so profusely endowed with sticks and stones it was inevitable that poking and pounding should ensue. Given human memory power, it was likewise inevitable that particularly handy sticks and stones should be retained, preserved, and treasured. Thus they become tools and the foundation of the tool-using culture of mankind. In recent years there has been much discussion of the supposed tool use of various animals. Many animals do indeed use various things in a tool-like manner. But no other animal retains and re-uses his "tool" and consequently nothing even faintly resembling the tool-using culture of mankind ever eventuates from any such behavior patterns.

Both science and industry are projections of this process. In a very real sense each is an aspect of the other. Hence, borrowing a figure of speech from John Dewey, I have been identifying science as the "thinking" aspect of the tool-using process and technology as the "doing" aspect of the same process. Neither is possible or conceivable except as an aspect of the other. Science advances through and with the advancement of the relevant apparatus (including the apparatus of mathematics) and *vice versa*.

So conceived, science is irrelevant to the whole universe-of-discourse of messiahship, and *vice versa*. But in common discourse in all languages we employ the term "science" to identify not only what scientists (or would-be scientists, or quasi-scientists, or mock-scientists) *do*, also what they *say*—with whatever degree of professional authority or intellectual justification. It was, of course, this aspect of "science" whose supposed "messiahship" I was discussing in this book. When I wrote, as the first "thesis to be nailed to the laboratory door": "That the truth of science is established only by belief, after the manner of all folk-lores," I was of course characterizing the literature produced by scientists.

From the beginning of tool-using and articulate speech men have romanced about their technological achievements. Indeed, such romancing has formed a considerable part of their folk-lore, and has therefore been a major impediment to the free play of what Veblen called "idle curiosity," and therefore of further technological advance. Thus each great technological revolution produced a vast outpouring of related folk-lore. The "miracle" of the first tools (including fire) sustained the animism which prevailed universally for aeons. The great agricultural revolution fostered the mythology of birth and death and rebirth which filled the earliest recorded literature.

Modern science and machine technology might have been expected to follow this analogy, in which case the literature of science would itself be the folk-lore of the present millennium. There is no inherent reason why it could not have been. Indeed, the eagerness of modern scientists and intellectuals generally to square the new ideas with the mythology of the past is one of the strangest anomalies of our intellectual history. But it is strange only *ex post*. *Ex ante* nothing could be more natural. After all, as late as the eighteenth century the scientists themselves were writing in a dead language. What broke the hold of ancient ceremonialism on modern culture, including the outpourings of scientists, was the cultural revolution: nationalism, capitalism, and secularism generally.

This was a gradual process, still obviously incomplete. Although secularism is clearly the order of the day, ancient traditions die slowly. There has been a considerable change during the half-century since *Science the False Messiah* was written. Not only is the New Science written in a different key from the old; like the new music it is atonal. We are told that the cosmology (if such it be!) of Planck, Einstein, Heisenberg and their co-workers cannot be set forth in the language of the multitude. It is totally irrelevant to the folk-lore of the past generations—and *vice versa*.

This is a consummation which Veblen would have understood and approved. The rise of scientific secularism may not be the triumph of the instinct of workmanship, but it is a step in the right direction.

HOLIER THAN THOU

One of the most fateful discoveries of modern science was the discovery of primitive society. What was discovered, or uncovered, of course was not the existence of what Rudyard Kipling called "lesser breeds," but the essential continuity of all cultures with each other. It is significant that the book which popularized the term "folkways" and brought the word "mores" into the English language (William Graham Sumner's *Folkways*) was published after the turn of the present century. The present generation of students who are quick to challenge each other's "value systems" find it hard to realize that such a challenge would have been meaningless to their grandparents.

It was perhaps in this sense more than any other that

The Theory of the Leisure Class was a pioneering book. Published half a decade before Sumner's *Folkways*, it was a true embodiment of the *Folkways* theory of value. All of Veblen's writings are replete with what today would be identified as "value judgments," and all his value judgments are derived explicitly from the folkways of "the establishment," as we say today. Thus he says of waste, in a celebrated passage:

> The use of the term "waste" is in one respect an unfortunate one. As used in the speech of everyday life the word carries an undertone of deprecation. It is used for want of a better term that will accurately describe the same range of motives and phenomena, and is not to be taken in an odious sense, as implying an illegitimate expenditure of human products or of human life. In the view of economic theory the expenditure in question is no more and no less legitimate than any other expenditure. It is called "waste" because the expenditure in question does not serve human life or human well-being on the whole...

The phrase "human life and human well-being" is especially significant, and I shall therefore come back to it shortly. At the moment my only point is Veblen's complete and seemingly unreserved commitment to the doctrine of mores-relativism, or moral agnosticism.

In this regard Veblen was a full quarter-century ahead of his contemporaries. By the middle of the century the triumph of mores-relativism was complete. Throughout all branches of social science virtually all professional scholars had accepted the new doctrine and made a great show of disavowing any moral absolute, identifying whatever "value judgments" they were aware of making as those of their age and generation. But there was an intermediate period when social scientists, though conversant with the new doctrine and well aware of the dominance of folkways and mores in all primitive societies still hesitated to apply this doctrine to themselves.

This little book, entitled *Holier Than Thou*, was written at that time, and its prevailing air of exasperation was directed at the then-older generation. While my teachers were still hesitating, I was plunging headlong into the new dogmatism. Perhaps this was a step that had to be taken. Moral agnosticism seems to be the essence of the folkways theory. That theory admits no absolute. If value judgments give effect to tribal practises *and nothing else*, then there are no absolute values. What we think good and right is what our community values. As human life is organized, it can be nothing more. There are no transcultural values.

Or are there? It was John Dewey's *Theory of Valuation* that "woke me from my dogmatic slumber." To be sure, nothing mitigates the obscurantism of "imbecile" institutions. But institutional archaism is not the whole of life. There is always the technological process, and the continuity of that process means that it is a locus of value no less definite than institutional taboos. The technological process is what Dewey called "a means-ends continuum." Human life itself is the "locus" of value—not in the animistic sense of totem and taboo but in the continuously progressive sense of the "instinct" (or process) of workmanship.

Veblen saw this. As I have suggested, even when he is intent upon pointing out the utter conventionality of "waste," he says in effect that human life and well-being is quite a different matter. Technologically considered, it is indeed. There is a hint of this in the closing pages of *Holier Than Thou*.

I could not have written my later books if I had not had the guidance of that hint.

INSTITUTIONALISM

Perhaps the briefest, clearest, and most definitive way to identify institutionalism is in terms of the two categories which are the concern of these two books. As a conception of the economy institutionalism differs decisively from the conception which has prevailed ever since the time of François Quesnay and Adam Smith. That, now commonly known as the classical conception, assumed the prevalence of private property and organized markets. Thus the economy so conceived is commonly identified as a market economy, an identification which glosses over the predominance of private property as well as the whole complex of cultural processes through which the present (and very recent) state of affairs has eventuated.

The founders of this system believed that the universe, including man, had been endowed at the Creation with a harmony of forces which needed only to be left alone to work to optimum effect. It was of course this belief which prompted Adam Smith to remark in his most famous passage that a man who seeks only the best bargain he can find "is led by a guiding hand to promote an end which is no part of his intention"—meaning, of course, the general welfare. Present day apologists no longer invoke "pre-established harmony." They argue only that, lacking any absolute truth or value, the market (for which read "the *status quo*") is the only alternative to chaos. Such negativism is less inspiring than Adam Smith's Deism, and its effect is no less stultifying.

Surely the species that has found its way from savagery to husbandry, and from husbandry to automation can do better than what Karl Marx called capitalism. The market system was a product of the industrial revolution of the eighteenth century. But we are now approaching the twenty-first century and a world-wide economy. Like the first stone hand-ax, the first fire brand and articulate speech itself, computerized automation is a manifestation of the technological process. Human life and well-being depends upon the furtherance of that process now no less than it did a thousand years ago when (as we have lately discovered) the foundations of the industrial economy was being laid, or a million years ago when mankind was first embarking upon its technological adventure. The values we seek are those of human life and well-being. The process by which we seek them is an experimental process, as it has always been. By pursuing this process we will go beyond capitalism, as our forebears went beyond the systems into which they were born.

This is the message of institutionalism.

University of Texas
January 1972

CLARENCE AYRES

SCIENCE
THE FALSE MESSIAH

CONTENTS

	PAGE
FOREWORD	11
PART ONE: THE BY-PRODUCTS OF INVENTION	17
I. Science: the New Folk-Lore	21
II. The Lust for Truth	31
III. Science as Invention	42
IV. The Lure of Machinery	56
V. A Kingdom of Machines	67
VI. Industrial Revolution	77
VII. The New Freedom	88
VIII. Dissolution	99
PART TWO: SCIENCE PRESENTS APOLOGIES	113
IX. The Mechanical Dispensation	119
X. Science Meek and Mild	129
XI. Making Religion Scientific	138
XII. Science unto Cæsar	148
XIII. Philosophy Embalms a Folk-Lore	161
XIV. The Reform of Superstition	175
XV. Science Betrays Religion	186
PART THREE: SCIENCE TAKES COMMAND	205
XVI. The Laws of Science	209
XVII. Seeing the Invisible Hand	220
XVIII. Society in the Light of Reason	231
XIX. Astrology and Pseudo-Science	242
XX. Science Will Provide	253
XXI. The Glory that is Art	264

CONTENTS—*Concluded*

		PAGE
XXII.	THE SCIENTIFIC MIND	274
XXIII.	THE RULE OF PSEUDO-SCIENCE	282
IN CONCLUSION: THESES TO BE NAILED TO THE LABORATORY DOOR		294

HOLIER THAN THOU

CONTENTS

CHAPTER		PAGE
I	PRE-VIEW	11
II	MAN THE ANIMAL	20
III	THE BASIS OF DECORUM	30
IV	O TEMPORA! O MORES!	45
V	GOOD TASTE	63
VI	FINE FEELINGS	79
VII	GOODNESS SANS SUPERSTITION	93
VIII	PIE IN THE SKY	108
IX	MORALS FOLLOW THE FLAG	123
X	GIDDY EMINENCE	137
XI	ONWARD AND UPWARD	155
XII	CONTROL BY EMULATION	178
XIII	THE RIFT IN RIGHTEOUSNESS	195
XIV	LAW VERSUS ORDER	127
XV	NO MORE CRUSADES	236

SCIENCE: THE FALSE MESSIAH

FOREWORD

PERHAPS the most depressing of all the aspects of our life within this vale of tears is the fact that nothing ever happens. Every generation has its hopes; and they never come off. Not only does Utopia remain unplucked; Nirvana is equally elusive. Only in books do civilizations ever fall. We may decline: that is a matter of interpretation. But only Messrs. Gibbon and Spengler have gone under. In the experience of peoples, things remain just about what they have always been.

In part this is because the span of human life is too brief for any man or generation ever to realize completely what may be actually happening. Moreover, there are always vestiges. Rome may have fallen. But the Holy Roman Empire disappeared—if it has actually disappeared—only the other day; while the still holier Roman Church seems to be endowed with immortal youth, a hardy perennial preserving to posterity at least the vestige of the authority of the Eternal City. We may talk of the decline of our ancient superstitions. That is the

SCIENCE: THE FALSE MESSIAH

proper thing to do in books. We may say that as a dominating force the folk-lore of Christianity is already doomed. Where is there to-day any man of power and position who not merely talks but acts with his eye on the judgment seat of the Almighty, or trembles at the threat of excommunication? Half a century ago Samuel Butler wrote that his countrymen would be as shocked to see any one practise as to hear any one doubt the tenets of their creed. It was not done. Nevertheless it continued to be said. As a manner of speech, as a part of the working idiom of the language, the folk-lore of the Hebrews is still a going concern and may continue so indefinitely. Power to make us tremble has passed, to be sure, into the hands of scientists. But the scientists, mindful of the *jus gentium* to the potency of which both the Roman and the British empires bear witness, have encouraged the people to continue the practise of their quaint customs, providing of course they do not act upon them. That old question of the Pharisees—to whom is our tribute really due?—continues to be pertinent; and the answer—to both—contains the wisdom of the ages. The past we have always with us.

Those, therefore, who expect the drama of industrial revolution to close with a holocaust and the doom of western civilization, are themselves doomed to bitter disappointment. Nothing as definitive as that is at all likely to occur. Such a thing may of course be invented. At some future time, some brilliant historian may achieve immortality by de-

FOREWORD

scribing the decline and fall of European—or medieval—culture; but as likely as not he will assign as the extreme limit of the civilization in which we live the year of "our Lord" 1901, because at that time an elderly lady died whose name had become symbolic, or perhaps 1914 because in that year an obscure Serbian killed an equally obscure archduke.

Of what use, then, is it to write books? The answer to this question is: none, except to the people who read books. People find it interesting to be told that science is the author of our greatest progress and our redeemer from barbarism and superstition. This is not true; but it is everywhere repeated. The same people, therefore, or some others, may find it interesting to be told that invention and technology are the basis of our new departures and the cause of our disaffection from our ancient folk-lore, and that science—far from redeeming us from the vices of credulity—is the inviolable faith of our new dispensation. Many people are edified by assurances that science accepts God and even perhaps the Holy Trinity. Consequently it is not unreasonable to suppose that some may be amused by the reflection that in doing so science cuts rather a poor figure. A great many people derive vast comfort from the transparency that science is competent to save us from the universal fate of decaying social systems and even to guide us by the power of reasoning into the haven of ultimate perfection. It may pain them to be told upon the testimony of the sciences that such is not the case. But it should at least prove interesting, since heresy is always livelier than dogma. The heresies may also be untrue. They may have no more truth in them than their corresponding articles of faith. But they may also have no less; and that should be enough to give them temporary currency.

PART ONE

The By-products of Invention

PART ONE

THE BY-PRODUCTS OF INVENTION

SUPERSTITION and conventionality—or as the pedants prefer to call them, folk-lore and folk-ways—are the Castor and Pollux of civilized society. This is rather strange, too, when we stop to consider it. Presumably the chief justification for the atrocities of civilization is the illumination of the highly civilized. They are our finest flower; and it is their peculiar distinction to be largely free from the conventions and superstitions of the multitude. Thus the truly civilized exhibit tolerance and breadth of mind, while the civilization on which such virtues flower is the accumulated refuse of ages of bigotry and provincialism. What, then, is civilization? To this question, the anthropologists unwaveringly answer: it is the process of accumulation of superstition and conventionality. We may take pride in its autumn flowers but the solid earth is established order and established belief. For the people as a whole, civilization means to believe in God, to honor the flag, and to respect the sanctities of private property.

To flowering humanitarians this view appears benighted, and our modern civilization is richly endowed with broad-minded men of such fine sensibilities. This suggests that we are viewing the autumn of an era. Sophistication and enlightened unbelief were general in Athens, in Rome, in prerevolutionary France; and those civilizations lamentably fell. The more humane among us, who ardently desire to raise the entire population to the level of cultivated gentlemen, would do well to consider these analogies. They would do well to consider who is most truly civilized, they—or the people who stay put.

Doubtless no one superstition is forever paramount in all the affairs of man. So far as we know, human civilization did not commence with a belief in God and Private Property; and we have no reason to suppose that it would come to an irretrievable end if such ideas should utterly disappear. No doubt this is a comfort to those who like to be remembered after they have gone. But evolution, and the universal flux of custom and belief, are a very different matter indeed from the supposition that civilization could ever be conducted, either now or in any future era of sublime enlightenment, wholly upon the basis of "reason" and with no admixture of tradition. We can not even talk about such a supposition and make sense. Whatever reason is, we can be sure of this: it is not a substitute for habits in individual life or for order in society.

Our own modern pretensions make this fact peculiarly significant. We ourselves are doubtless

THE BY-PRODUCTS OF INVENTION

passing through a period of enlightenment, as such things are called. Beliefs are increasingly in disfavor among us, and superstition is an epithet of opprobrium. We are aware of what is happening. We are aware also of the interpretation that has so often been put upon this phenomenon. But we are not particularly disturbed. Our case, we feel, is different; and the reason for that difference is—science. As we like to say, science makes all the difference!

But we hardly realize, when we talk after this fashion, that we are only exhibiting how little we actually know about this great emancipator of our dreams. If we would only reflect upon it, our very faith would show us that science is the object of our devout belief. It is the great constant, superstition, in another guise. If we would do more—if we would inquire into its antecedents—we should soon discover further that the prime mover in our recent developments is not that galaxy of noble truths which we call science, but the thoroughly mundane and immensely potent driving force of mechanical technology. Science is the handsome Doctor Jekyll; machinery is Mr. Hyde—powerful and rather sinister. Science is the Pentateuch of technology— what we have been given to believe by our new machine-made folkways. We may even notice, with something of a start, that the new Story of Creation has not come in answer to the crying of our hearts; it is not an outgrowth of our age-long experience. It is a product of the machine-shop, and it pro-

[19]

SCIENCE: THE FALSE MESSIAH

nounces its blessings there and nowhere else. Our ancient culture showed us how to live after a certain fashion. The new technology, and our new superstition, teach us only how to work. Machinery is our Babylonian exile.

CHAPTER I

SCIENCE: THE NEW FOLK-LORE

SCIENCE—so we have been constrained to think—is true. All its major propositions have been definitely established to our complete satisfaction. This is what makes it so hard to think about. Only with the greatest reluctance do we ever seriously reflect upon what we have already decided to believe, or inquire in retrospect how we came to believe in it and just what there is about it that has so completely captured our assent. Because science is true, we omit to inquire what the limitations of its truth may be. Because the truths of science have been established, we lose interest in how they were established. Indeed, we do not wholly admire the type of mind which insists always upon poring over limitations and short-comings instead of accepting all blessings with a thankful heart and a happily sated imagination. To doubt any accepted belief is of course impiety, a thing which we all righteously abhor. It even verges upon impiety to make such inquiries as might lead to doubt. This is pretty generally our feeling about science, and a very significant frame of mind it is. It suggests, by direct

SCIENCE: THE FALSE MESSIAH

and forcible comparison, that science has attained the state of an established creed.

To be sure, science does not represent itself as folk-lore. But then neither did the folk-lore of our superstitious past. Folk-lore never does. We must not imagine Moses coming down from Mount Sinai and urging Joshua and Aaron to bear in mind that his various narratives are folk-lore. It was enough that they were marvelous. Joshua and Aaron understood them so, and no doubt interpreted them as accurately as we interpret our neighbor's stories of the voices he has heard over his radio coming out of a thick cloud of static. It is a mistake to think that Joshua and Aaron thought them any stranger than that. But it would also be a mistake to suppose that the Israelites were as surprised by Moses' story as we should be, or as surprised as they would have been to hear him say that he had been borne through the clouds at one hundred twenty miles to the hour and accompanied by the sound of an awful roaring. Sufficient unto the day is the folk-lore thereof. In Moses' time direct communion with the Lord was uncommon, but by no means unknown. His miracles were of the same order as those performed by all the contemporary prophets. So are ours.

The proper parallel to the miracles of ancient prophets is, however, not the commonplace adventures of ordinary citizens but the accomplishments of our own prophets, the men who have had direct communication not with the Jahve of the Israelites but with atoms and electrons, with stars so far off that their distance is measured in the number of years it takes their glimmer to arrive, with dried-up oceans and glaciers before the dawn of history, with the gorgons and chimeras that inhabited the earth before those glaciers came, with animalculæ so small that they pass through the finest filter and so powerful that they will strike down a man in a few hours. These men tell tales of the creation of all living things from primordial ooze, of the origin of the earth from spouts of incandescent gas from the sun, of rays that penetrate the solidest seeming stuff, of the electron-stuff of which all things are composed. They sing of matter which is not matter but energy, of warped and curving space, of fixed points which change places from moment to moment, and of different moments which are simultaneous in different locations. These are the real marvels of the age of science. We must not dismiss them lightly because we believe that they are true. After all, the Israelites believed Moses. That is how they can help us to understand the character of folk-lore.

Moreover, these modern tales of the creation and composition and regulation of the universe provide another clue to folk-lore: they imply vast mysterious forces in the background, sublime powers at which they can only hint. They are the powers which rule our lives. Ordinary men can not come nigh them. They can be approached only by the prophets, and even by them only at appointed spots and through the invocations appropriate to whatever powers

SCIENCE: THE FALSE MESSIAH

they be. Such, always, is the meaning of the legends.

We must not be misled by our belief that the folk-lore of our time has all been "proved," and so is not lore at all but fact. This, also, is the nature of folk-lore. The legends of creation can always be proved, each in its own appropriate way. Every time a woman feels the pains of child-birth, she proves the truth of the Hebrew legend of creation. After the Fall, Jehovah said: "In sorrow thou shalt bring forth children." And so it is. The obvious objection to this demonstration of the legend is that although, if we assume the legend to be true, we find the current facts bearing it out, we must also grant that if we assume the legend to be false, these facts may be explained by some other legend even better—for example, by the legend of evolution, according to which we suppose that the difficulties of parturition are due to the effect of our erect posture. But what we overlook when we urge objections like this against other people's legends, is that the same objections can be urged against ours. The theory of evolution is a case in point. No doubt we possess an enormous number of facts concerning the resemblances between the structure of our bodies and that of anthropoid apes, and back of them the mammals generally, and back of them vertebrates generally, and so on. The theory of evolution "fits" these facts beautifully, so we say; and it is true. But the legend of the Fall fits the pains of child-birth beautifully. In neither

SCIENCE: THE NEW FOLK-LORE

case does the legend do any more than fit. The facts are what they are; and rather more so: there are always more facts which do not fit the theories, and about which we do not hear so much—until a new theory has been invented into which those erring facts do fit. No fact ever obliges any one to invent a theory, or to believe one theory and not another. People believe theories and legends and all sorts of folk-lore for other reasons.

The leading reason is that they are accustomed to it, so that it sounds perfectly natural, and therefore plausible. A folk-lore does not spring into existence over night, nor even during forty days and forty nights. At the end of such a period, legend has it, Moses emerged from Mount Sinai with an extraordinary code of public law and divine etiquette. But even if we accept this legend at face value, we must still allow that the Commandments and all their corollaries, running through the Book of Exodus and most of the Book of Leviticus, form only a portion of the folk-lore of the people. Moses did not invent Jehovah with all his functions as tutelary deity of the Israelites. All this was prepared for him. Consequently he could claim little originality for the circumstances surrounding the revelation of the tables of the laws: granted Genesis and earlier portions of Exodus, the episodes upon Mount Sinai were not only prepared for; they were inevitable. While if we take a later point of view, and regard the Sinaic legends with a critical eye, it becomes extremely probable that so extensive

SCIENCE: THE FALSE MESSIAH

a code as the Mosaic one must have developed gradually, becoming established bit by bit after the manner of the folkways, and only becoming the subject of a legend after the event. The Mosaic folk-lore thus appears as a codification of a much more extensive body of folk-lore of which it was a free translation.

All this, too, is perfectly illustrated and illuminated by science. Our scientific lore is decidedly extensive—vastly more extensive than the legendary history of the Israelites. But it has been a very gradual growth. No major hypothesis has ever emerged fully formed from the brain of a single inspired prophet. The prophets have always built upon the familiar and the accepted in large part. Such a miracle as Einstein's relativity is sometimes spoken of as "revolutionary." But any one who cares to look into that admirable edition of Poincaré's *Foundations of Science* to which Josiah Royce wrote an introduction, will see that the philosophical and mathematical basis of relativity had been gathering for several decades, while the famous Michelson-Morley experiment upon the velocity of light (the experimental "fact" on which relativity was based) can be regarded for the purposes of this legend as the voice of Jehovah speaking out of a thick black cloud, calling Einstein up upon the mountain. Evolution, another "revolutionary" conception, which completely "reorganized" biology, was also a gradual accretion of ideas which had already become traditional before the revelations of Charles

[26]

SCIENCE: THE NEW FOLK-LORE

Darwin. Darwin's relation to evolution is quite similar to Moses' relation to ancient Hebrew lore: it is associated by tradition with his name.

The authority of these names is very great. We must allow that Darwin is a name to conjure with; and this helps us to understand the attitude of the Israelites and the later inheritors of their folk-lore toward Moses. But it is an equivocal authority. If we say that evolution is not legendary but true, and cite the authority of Darwin for the justification of our belief, we fall at once into a predicament, should any one require us to explain the authority of Darwin. Darwin's greatness is established by the same folk-lore that establishes evolution. We can hardly say that evolution is true because Darwin discovered it, and then say that Darwin is reliable because he discovered evolution. We do say exactly that, however; and this further illuminates the character of folk-lore. That is precisely what we find in the case of Moses. According to his own story, the Commandments are to be accepted on his authority; while his authority is vouched for by the fact that to him alone were the Commandments thus revealed.

A great deal has been said, first and last, upon the subject of authority. The defenders of the faith of science have been at considerable pains to expose the weakness of authority as the basis of a folk-lore. In all this they have been quite right. Their only mistake has been their naive belief that all other folk-lores except their own rest only upon authority; whereas the truth of the matter is that no

SCIENCE: THE FALSE MESSIAH

folk-lore does so. But it is one of the rules of the game that the other fellow's legends are supported by no other prop than the credit of their legendary heroes. Elijah considered that he had only to discredit Baal to discredit all his followers, and Elijah used for the purpose a test of his own devising. Scientists do the same. Holding that the folk-lore of a church is established only upon the authority of the legendary authors, they have in many cases discredited those authors quite completely—according to tests that are accepted among scientists. This is like discrediting physics by denying that Newton was the discoverer of gravitation.

In a delightful little essay on certainty in science and theology, first published some years ago in the Hibbert Journal, Professor Charles Cobb, a mathematician, made an interesting point. Like science, he said, folk-lore is founded upon axioms. The truth and authority of each depend entirely upon the axioms. If its axioms are true, the rest of the body of theory or legend which is derived from them is true also. This does not seem at first to be much of a point. The devout believers in any folk-lore are quite likely to accept it, but with a reservation. "Ah," they will say, "but those axioms! There is just the difference between our knowledge and the false theories of unbelievers." But so far as science is concerned, there is nothing to be said for the axioms except that they are what we are accustomed to believe. This is the authority upon which we receive the initial axioms of geometry. "Let us suppose what is only reasonable, as follows: that a straight line is a line such that if any part of it be cut off and laid along any other part, so that the ends of the first part lie upon the second part, the two parts will coincide throughout and be one and the same straight line." This seems reasonable because we are accustomed to sighting. Our most familiar imagery on the subject is the straight edge of a board along which we sight. If, when we sight from one end to the other, all the intervening edge lies directly in the line of vision, we consider the edge straight. The axiom only defines this process, somewhat laboriously. The upshot of it is that a straight line is the path of light. But if we define a straight line directly as the path of light, and if Einstein is right about the path of light in a gravitational field, then surprising things follow—not at all such as we should expect from a perusal of the elementary propositions of plane geometry. And there is no reason whatever for preferring one definition to another, except that it is what we are accustomed to.

The nether part of the argument requires no explanation. Few folk-lores expose their axioms to view as nakedly as mathematical assumptions. But it is comparatively easy to see that other folk-lores are nevertheless based on assumption in very much the same way. And if Judaism did not list its axioms, neither does biology.

If anything is lacking to complete the picture, it is the awe in which we hold the truths of science and the prophets to whom we attribute these revelations.

SCIENCE: THE FALSE MESSIAH

We say, of course, that we stand in awe before the awful truth, and that we venerate the saints of science because of what they have done for humanity in giving it the truth. If we ever notice the passion for canonization by which so many scientists are fired, we set it down to a laudable zeal for the welfare of mankind. This, again, is precisely the state of mind of every loyal believer. However primitive his folk-lore, no one ever stands in awe of what he does not credit, or reverence men whom he considers humbugs. Our attitude toward science is the last perfection of unreason, the final genuflection of the faithful, as our explanation of it makes amply evident. We honor because we believe; we believe because we honor. The lore giveth and the lore taketh away; blessed be the name of the lore. Folk-lore becomes holy by the same process by which the *mores* become holy: because the folk-lore makes it so. Folk-lore is a body of truth verified by repetition and sanctified by faith. This proposition includes science.

CHAPTER II

THE LUST FOR TRUTH

MEN who carry on a civilization are under the necessity of meeting every sort of exigency that can arise in a universe like this and in the lives of men like us. The theory of numbers, the quantum theory, the relativity of time and space, the planetesimal hypothesis are remarkably interesting and consistent. But as a complete intellectual equipment for a man who has got to make a living and marry a wife and bring up a family and vote and hold office and assign his fortune to his heirs in perpetual trust guarding against the follies of his descendants unto the fourth generation, even the differential calculus, potent as it is in number work, is unfortunately insufficient! In the complete furnishing of the mind of every man, common and stupid or sublimely sophisticated, non-scientific matter of fact and fancy out-bulks the cameo of science by so much that the two are incommensurable.

This does not mean that the importance of science is overrated. That would be impossible. But it does mean that science is important not as an intellectual content upon which a civilization can be built, but for other and quite different reasons.

SCIENCE: THE FALSE MESSIAH

The ideas upon which civilization turns are homely ideas, affecting all the ordinary things that men do. They are suppositions about life, and man, and the physical world not as it looks through a microscope or a telescope but at its point of impact upon the ordinary affairs of men. These suppositions rule life, and to them science is as irrelevant as the lost literature of the Minoans. For that matter, science is couched in a language as difficult to acquire and as remote from possible employment by ordinary men as a dead language. Science can not be the intellectual background of the common man because science is not intelligible to the common man, and never will be—barring the realization of that Shavian dream of a new race of men who will have the remarkable faculty of passing on the higher mathematics to their progeny like a family resemblance. But whatever the difficulty, ordinary men are not going to make any special effort to master the calculus until and unless it proves as necessary and as important for all the common affairs of life as a knowledge of the written language has now become. That time is not yet. Very few highly educated men now understand it, for the same reason that very few highly educated men now understand Siamese—because it does not affect them directly even in their thinking.

Our habit is always to think of scientific researches as contributions to "knowledge," to which industrial application is quite incidental. But to what knowledge? Not to common knowledge, certainly. Science is never common knowledge, because it never becomes either the common possession or the common reliance of men in dealing with their common affairs. Science is a contribution to knowledge only if we mean by knowledge, science. Piling up more and more science does not make the human race wiser and wiser in all its ordinary affairs. It simply piles up more and more science, each bit added to the pile being as remote as the preceding bit. If the impact of that research upon common men is transmitted only by machinery, it makes no difference that the more machinery we have the more machinery we get. That is true of course. Scientific researches have an immediate effect upon ordinary life; but only through the rapid adaptation to common use of the inventions so engendered. They have in addition the secondary effect of building up more science and thus making further mechanical adaptations possible. This has the tertiary effect of building up still more science and making still more mechanics possible. Portentous vista!

The mystery is what motive rules the universities that they should provide laboratories for the tool-making industry. Commerce is more transparent. The object of commerce is to find and exploit the sure thing. Individual gold-diggers take chances; but the value of gold and the lucrative character of gold-digging as a business is the certainty about which individual fortunes describe their orbits. Commerce has only been waiting for

SCIENCE: THE FALSE MESSIAH

science to prove its character. In earlier days, the researches of pure scientists must have seemed very dubious indeed, even to the shrewdest Shylock. But that was only the formative or probationary period of science. Furthermore, though commerce existed in that day and pecuniary motives throve, manufacturing did not. The manufacturing establishment came into existence at the same time with science. The "applications" of science and the exploitation of the applications of science coincided. At first, both were sporadic and individual and accidental. Research waited upon the whims of individual enthusiasts. Invention and manufacture waited upon the ingenuity of hardy innovators. The organization of research councils and chemical and electrical industries is a late development; but its arrival has been marked by the convergence of manufacture and research. This convergence is quite inevitable. One great electrical concern which is said to control the market for electrical appliances of all kinds has for some time published a series of "good will" advertisements in which it describes itself as a firm that will profit by any conceivable achievement in the field of electricity. If this firm does control the industry, this boast is obviously just. It will exploit whatever is turned up in the researches of the scientists. Its well-known hospitality to research is perfectly logical and inevitable. Its laboratories are directed by a man of great scientific eminence, and he is said to enjoy as free a hand for even the most esoteric researches

THE LUST FOR TRUTH

in "pure" science as ever Curie had in that abandoned shed where he first isolated radium. Meanwhile the Chemical Foundation—an organization through which the great powers of the American chemical industry express themselves—is rapidly buying up the chemical laboratories of all the universities by putting both professors and students upon the pay-roll of the industry. When all is said and done, the business men have not been slow about taking over the goose that lays the golden eggs once it reached maturity and began to lay in paying quantities.

Thus the motives which lead business men to endow research are clear enough. It is the universities that remain obscure. Why should universities maintain scientific laboratories at all? To be sure, the individual scientists continue to disclaim pecuniary gain. They are interested, so they say, not in mere vulgar inventions and their applications, but in scientific "truth." But what does scientific truth amount to, that universities should take an interest in it? The business men are of the opinion that it chiefly amounts to more inventions, and they are ready to back their hunch with real money. In this the scientists agree with them. Each scientist of course maintains the "purity" of his own research and repudiates violently any proposal to subject his pet research to the test of immediate applicability and "usefulness." It is not the object of his research to be useful but to be sound. Nevertheless, no one is more insistent upon the ultimate

SCIENCE: THE FALSE MESSIAH

usefulness of all research than the scientist. One set of data or one bit of theory may stand forever uncontaminated by the herd. But beyond those fragments—made possible by them—grand discoveries emerge by which "all humanity is enriched." Science is the fairy god-mother of civilization. This is the fashion in which scientists talk. What they mean is that they are individually chaste, but science is common to all.

One might infer from this that universities should endow scientists because they are pure; but that science, since it is always so profitable, might well be taken over by firms which specialize in profits. This is a paradox. But perhaps it means only that scientists distrust Moloch as a taskmaster. They admit freely that their researches will ultimately be profitable; but they hesitate to have them conducted intentionally for profit lest the test be applied short-sightedly. Yet if science is so profitable in the long run, if they think so and the business men think so, and each needs the other in his business, why should the scientists so distrust their ability to enlighten the manufacturer in a matter so near to his heart and dear to his purse?

At this point, the motives that govern individual scientists can be reduced to venality. When academic honors are on the wind, scientists are as keen upon the scent as any theologian. But why should academic honors continue to be available to scientists? Because, apparently, the universities persist in treating science not as invention but as folk-lore.

THE LUST FOR TRUTH

Each new hypothesis excites the equal though opposite enthusiasms of financiers and faculties. The financiers see the assurance of future gain. The faculties see knowledge: a new acquisition to the precious store of demonstrated truth.

This lust of universities for scientific truth would seem a strange thing if it were not so familiar. Why should universities care to know the truth about electrons, and pterodactyls, Betelgeuse, and the path of light in a gravitational field? We have heard of the thirst for knowledge. Universities cultivate it. But other thirsts are frequently overdone and lead to drunkenness. The mere existence of a thirst is not ordinarily regarded as sufficient reason for gratifying it. Indeed, quite the contrary is the case. Morality has often been described as the suppression of thirsts in the interest of reason. Not even a thirst for knowledge is moral unless there is some good reason for possessing the knowledge that is thirsted after. A passion for knowledge without discrimination is as mad as a passion for the moon. Such a mania should appropriately end in suicide in a fit of despondency over our utter hopeless inability ever to take observations of the far side of the loony satellite.

Once upon a time cosmological interest was quite sensible. At the time of Copernicus, for instance, the cosmos was generally supposed to be arranged according to a divine pattern in which the fate of mankind could be read. Then astronomy was a humanistic science. Chemistry was likewise

SCIENCE: THE FALSE MESSIAH

a humanistic science when it held out some reasonable promise of the elixir of life, the philosopher's stone, and other solutions to the riddle of existence. Men of liberal education, whose interests were broadly humanistic, were then impelled to take up the study of these subjects in order to inform themselves concerning matters that lie nearest to the human heart. In other words, when chemistry was alchemy and astronomy was astrology they were not technical departments and gave rise to few technical innovations; but they were humanistic. They had something to teach that was worth any man's while to know. It may not have been sound, judged by the mechanical standards of the present day; but it was relevant.

All this has long since passed away. If science had stopped short with Copernicus, a readjustment would no doubt have soon been made. The solar system would have been assimilated to the prevailing folk-lore so that meanings would have been read in the positions of the planets, just as farmers still read meanings in the phases of the moon. Under such conditions, it would have been necessary for every man to understand Copernican astronomy. But science did not stop. On the contrary, it has moved faster and faster ever since that time, with the consequence that it has lost all human significance and has become a completely barren technology. We have only to compare science with the folk-lore which surrounds the rising and setting of the sun, and the passage of the seasons—simple natural occurrences which science can never wholly steal away from the eyes and hearts of ordinary men—to see why sunsets will always be important in human life and why electrons can never be. No poetry can be written about "the smallest known component of matter"! Morality is never likely to take cognizance of the magnitude of Betelgeuse!

Indeed, what could be more tell-tale than the attitude of poets, and artists, novelists, theologians, moralists, politicians, musicians, and men of genius generally toward this tremendous body of knowledge with which the universities are crammed? What poet or musician or statesman knows or cares the tiniest iota for any single fact of modern science? They may approve of it. They may be glad of it as a salubrious feature of the common wealth. They may take pains to employ competent doctors and automobile mechanics. But do they care to know these awful facts in the presence of which universities are said to tremble? Not in the least. An eminent and brilliant physiologist has recently complained that no English poet since Keats had even a decent working knowledge of science. This implies that Keats had, which is not true. Keats was a medico's apprentice, and had the knowledge of science appropriate to that station in life. No great English poet has ever had a working knowledge of science. The physiologist, Mr. J. B. S. Haldane, bemoans this fact, and threatens us with a falling off of poetry until it is rectified. In that case we have seen the end of poetry. What could a

SCIENCE: THE FALSE MESSIAH

poet get from physiology? An ode to the pituitary gland!

And yet—why not? Rembrandt painted a picture of an anatomy lesson, and more recently D. H. Lawrence and Sherwood Anderson have taken all the characters for their novels from the case histories of psychiatric clinics. A great physicist has assured us that the colors to be seen in the spectroscope are very beautiful. A mathematician once said that Napier's analogies (a set of propositions in trigonometry) are a consummate work of art. Lecturers in astronomy always explain to the ladies of their audiences that the earth could be dropped into a sun-spot without making a puff of smoke visible at this distance, and the ladies are always interested. Sun-spots and microbes and electrons have definitely arrived. They are just as cultural to-day as the hypothetical amours of Keats and Robert Louis Stevenson. These are esthetic satisfactions!

But this is not science; it is folk-lore. Science has become the most extensive and intricate body of "knowledge" ever got together by the mind of man—extraordinary, imposing, respect-compelling. This body of lore can be treated in two ways. It can be made use of as an instrument and a method for producing certain effects. Such use requires meticulous familiarity with its details and the most delicate precision in its manipulation, and is utterly inaccessible to all but those few who make it their profession. It has, moreover, no interest for any but professional scientists. But the lore of science, nevertheless, has another interest. It is interesting because it is inaccessible, because it is obscure, because it is mysterious, because it is esoteric and unintelligible. Nothing mysterious and esoteric ever fails to attract some attention.

Science has attracted great attention because it is also, in ways we do not clearly understand, extravagantly potent. Not only are its doctrines couched in the most delightfully exotic language—who could fail to thrill to "enzymes" and "trypanosomes"?—but these things actually mean something! We are compelled to respect those meanings, and that genuine respect is the foundation of our interest. We must never forget that what places such a thing as relativity on the market as a merchantable mystery is the fact that by its conjuries men have really predicted and verified the dislocation of certain extremely distant stars. No one cares about these stars in any other connection. They have no portents for the guidance of human life. But we overlook this. The fact remains that their location was a prodigious feat. It was accomplished by science. Therefore, we feel, it behooves us to know something about this science, just as it behooves us to know something about Shakespeare and the Imagists. Though our sins be as scarlet, we shall speak with easy familiarity of filterable viruses and the quantum theory. Such is culture; and such is the contribution of science to common knowledge.

CHAPTER III

SCIENCE AS INVENTION

WHEN Moses emerged from the cloudy obscurity of Mount Sinai and stood before the people with the stone tablets in his hand, he announced that his laws were based on direct observation. It is not recorded that any one doubted him. He had gone up to the awful places of the Most High. That also was a fact of general observation. That he had seen there whatever was to be seen was the only reasonable inference.

The credit of science as a body of folk-lore rests upon the same foundations. Scientists emerge from the awful obscurity of their laboratories and announce that these, their decalogues of physics and biology, are based upon direct observation. We do not doubt them, because they have been in the laboratories and we have not. Whatever is to be seen there, they have seen: electrons, microbes, and spiral nebulæ. Their lieutenants, who have been with them in the place of mystery, as Joshua went with Moses and "abode in the mountain" and "went not out of the tabernacle," bear them out in every particular, precisely as Joshua bore out Moses. Closely pressed, the scientist will admit that he has not seen the electron face to face: he has devised

SCIENCE AS INVENTION

a delicate instrument in which fine drops of oil in a vacuum are mounted by electrons and exhibit their presence and potency. Those most distant stars, also—they have been observed upon photographic plates, many, many times enlarged. But not even Moses claimed to have looked upon the full face of the Almighty; the Lord lifted His hand, and Moses saw His back parts going by. Scientists look upon the back parts of electrons.

In making their case for science, the modern prophets appeal directly to our credulity precisely as Moses did. They may not be conscious of the fact. But then, it is not recorded that Moses was. We have no record of Moses saying to Joshua, "Come, let us make a legend for the people, that they may believe to their greater good." Moses declared: "Thus saith the Lord," and no doubt meant what he said. Scientists declare: "Such are the facts," and certainly mean what they say. But the expression "fact" is an appeal to traditional belief, exactly like the expression "Lord." A "fact" is a conception which people agree upon; that is, a tradition; that is, folk-lore. "It is a fact that it is five o'clock. Your watch agrees with mine. We both regulate our chronometers by Naval Observatory time. By so doing we allow ourselves to be guided by international agreement—and how can we do otherwise?" We sometimes say that we are guided by the stars; but what really rules us is our beliefs, or international agreements, about the meaning of the stars.

SCIENCE: THE FALSE MESSIAH

When scientists appeal to "facts," they mean to rest their case on our belief in microscopes. That is why they utter the appealing words with such a show of emotion. With shining eyes and heaving breasts they pronounce their solemn invocation: "Facts; based on direct observation; quantitatively recorded; with mathematical accuracy." This means: established tradition; based on general agreement; in the use of machinery; calibrated according to the series of cardinal numbers. The number system is of course conventional. The readings on the dials of instruments of "observation" are also conventional. The employment of those instruments is similarly conventional. So also is the treatment by which the readings are arranged in orderly theories, or legends. In fine, laboratories differ from kitchens and drawing rooms only by being rather more conventional. The etiquette of science is stricter than that of society.

What particularly distinguishes science is its use of instruments. To hear some scientists talk, one would think that scientists are the only ones who have ever used their eyes. This is not literally true. The men of science have been profoundly wise, careful and industrious; but they have not been so much wiser or more industrious than all the rest of the population as the difference between the "order and precision" of science and the "confusion" of "popular thought" would seem to indicate. That is incredible. It is incredible to suppose that all the wise and industrious men of

SCIENCE AS INVENTION

modern times have gone into science with never a one left over to bring order and precision into "popular thought." Our imaginations boggle at so monstrous an assumption. It must be that the order and precision of science are a peculiar kind of regularity, appropriate and attainable in science, and most likely inappropriate and unattainable elsewhere.

The clue to the particular nature of scientific observation, and accuracy, is mathematics. Scientific observations are mathematical. Now how can mathematical observations be made? Is it possible for a very superior intellect, by sheer force of superiority, to make a mathematical observation upon the value of life, or the holiness of monogamy? We have some evidence upon this question in the fact that no one ever has. But an examination of the process by which scientists do make their exact statements indicates that no one ever will. It is frequently said that these are not fit "material" for scientific study. That raises the question, what is fit material?—and suggests the answer that science can only deal exactly with material susceptible to quantitative observation. Thus the whole problem runs back to the nature of quantitative examination.

Fortunately this is a process with which we are all familiar. The use of a tape measure, a thermometer, a T square, a gas meter, a hydrometer, an ammeter,—all these are methods of quantitative observation. In short, the greatest common denominator of accurate observation is the meter, of

SCIENCE: THE FALSE MESSIAH

whatever sort: the machine devised to record in successive, equal, conventionally numbered units the increase or decrease of any volume, length, width or degree of any kind. Whenever and wherever a happening can be trained through a machine, and that machine converged upon a dial, and that dial marked off into standard units, and those units numbered: then and there an exact, scientific, mathematical observation has been made possible. In some cases this is very easy. To a naked savage it might seem quite beyond the intellectual powers of man to record the varying speeds of a vehicle with numerical accuracy. But to any one who knows the principle of the mechanical governor, it is the height of simplicity. A governor is geared to a moving part. The free end of the turning governor is geared to a finger moving across a dial, or a cylindrical dial moving across a stationary finger. The position of the finger on the dial, then, at once indicates the precise speed of the turning parts. To make the reading scientific, it is then necessary only to number the dial upon the approved, conventional, universal standard—and the scientific observation is complete. The mathematics is merely a language of numbers, a very complicated language, to be sure, to which the word "merely" does scant justice; but it is a mere language in the sense that no observations can be inferred from it. No science can be deduced from it—except further complicated phrases in the language of numbers. The deductions or inductions of science are made by

SCIENCE AS INVENTION

the mathematical statement of "readings." A moment's serious reflection will show any one that this is true. Modern science, as distinguished from other bodies of folk-lore treating the sun and the stars, man and the elements, springs from just one source: that is, from instruments of precision. That is, from machines.

The history of science does not exhibit this source clearly. But then, history does not exhibit anything clearly. History, we might say with a grimace at our own paradox, is not science, even when it is recited by scientists and is the history of science. The truths of modern physics, and the accounts of the history of modern physics, are two quite distinct bodies of knowledge or belief. Our common belief is that physics sprang full-grown from the intellectual brows of Newton, Faraday, Helmholz, and the rest. Physics is a body of truth, and therefore presumably came from the minds of thinkers. This seems reasonable, and appeals to our prejudices. It makes the great achievement of science a personal triumph of a series of great men, and so it glorifies the human intellect. It makes the development of an astoundingly complex and mysterious system of axioms, formulas, observations, and hypotheses a simple and accountable affair. Asked by our children where science came from, we have only to say that it came from Copernicus, and Harvey, and Mendeljeff, and Pasteur, and let it go at that.

But history of this kind is nothing more nor less

SCIENCE: THE FALSE MESSIAH

than a series of miracles. Such a tradition requires us to believe in the magical genius of the Koperniks and the Metchnikoffs. We are perfectly willing to do this, of course, because it is a tradition. We are well accustomed to the same tradition in all other history. The history of nations is likewise a series of miracles performed by Napoleon, Wellington, Washington, and the other fathers of their countries. The first history we ever learned was the names of the Presidents: Washington, Adams, Jefferson, Madison, and so on; and we learned it with a clear understanding that these men by their successive fiats had brought about the United States of America as at present constituted. To repeat Copernicus, Galileo, Newton, Boyle, only continues the tradition, and assigns to science the same magic authorship to which we have previously assigned civilization. If we later correct our understanding of social history and discount somewhat the exclusive influence of Adams and Madison, we have the same discounts to allow in the history of science—though we seldom get that far in our revisions.

In short, the history of science is shrouded in the same obscurity that covers all parts of civilization. As we optimistically say, it has not yet been written. But if we can nevertheless see that natural resources played a large, though uncelebrated, part in our national history, we can imagine at least the part that may have been played in the development of science by machinery and instruments. For example, a certain period reveals a general eruption of

SCIENCE AS INVENTION

discoveries in a certain region of science. One way to treat this occurrence is to pick out a prophet and assign everything to his "influence." Another is to inquire what the methods were that were employed in all these researches. As a result of that inquiry, it may instantly appear that the compound microscope had just come into use as a product and development of the lens-grinding industry. All these discoveries were the observation of one thing after another with the compound microscope. As a matter of fact, the compound microscope, like many other instruments of supreme importance to science, was actually devised by a man who was more interested in microscopes than in what could be seen with them. But many, possibly most, of the inventions by which science has been advanced, have been made by scientists themselves. The electron illustrates this perfectly. Sir William Crookes, so the story goes, was possessed of an almost legendary skill at glass-blowing. The joints of his glass apparatus were articulated with uncanny art. He was therefore able to obtain rarer vacuums than had ever been produced before, and to observe in them many various performances which can be staged in vacuums. One of these—it takes no great stretch of the imagination to suppose such a thing—was the performance of an electric arc in a vacuum. As every school child knows, the ordinary electric arc employs the air, or the water and electrolyte, as a vehicle of the current. In a perfect vacuum this vehicle would be removed. It

SCIENCE: THE FALSE MESSIAH

would therefore be interesting to see what would happen. Crookes tried it, and saw that what does happen is an electric ray extending in a straight line from negative to positive pole, and producing an intriguingly fluorescent area in the vacuum tube. The imagination of the physicists was instantly captivated. Every one, more or less, began to monkey with Crookes' tubes, as they were hereinafter and forever called, and various other performances appeared forthwith. It was discovered, for instance, that the mere presence of a magnet deflected Crookes' "cathode rays." It was discovered further that, when the cathode rays strike any body, they give off from the point of impact another kind of ray, itself invisible, which has the strange faculty of "exposing" photographic plates and, second and most remarkable, of passing through opaque substances to do it. Thus the German physicist, Roentgen, who was interested in photography before he took up with Crookes' tubes, had photographic plates lying about, suffered the misfortune of having them "light-struck" by the invisible rays, observed that the rays passed through different substances with varying ease, and so was able to announce to the world the discovery of Roentgen rays (or X-rays).

Meanwhile, Sir J. J. Thomson (not yet dignified by knighthood) followed up the suggestions of the magnet's deflection of Crookes' cathode rays. Suppose, he thought, that these rays are streams of minute particles of matter broken off from the

SCIENCE AS INVENTION

negative (cathode) pole to carry the current of electricity across Crookes' vacuum. By taking account of the strength of the magnet and the amount of the deflection it will be possible to calculate the size of the particles. He made this calculation, and emerged triumphantly with the first demonstration of the size of the electron, here displayed for the first time on any stage as the smaller constituent of atoms.

Then, somewhat later, a young physicist named Moseley—he was still in the flush of youth when the Great War claimed him—perfected a novel spectroscope capable of showing the spectra of Roentgen rays. By this combination of machines, and by the aid of calculations involving certain previous measurements, the magnitude of light waves of different colors, and so on, he was able to rearrange and correct the "periodic table" of the elements which is the foundation of all chemistry, and thus to set in motion the reorganization of that whole science. Such are some of the effects staged within Crookes' vacuum tube: a triumph for glass-blowing!

But Sir William Crookes was not an Edison. He was not even interested in the romance of invention, as a technologist or engineer might be. On the contrary, his was a theoretical mind. This distinguishes him as a scientist. In his apprehension, an instrument, however delicate or revolutionary, was a means to a scientific end. Nevertheless, to gather facts, and on the basis of those facts to formulate

SCIENCE: THE FALSE MESSIAH

new theories, it was necessary for him to invent a machine; and he accordingly did so. Moreover, having attained his theoretical results he did not destroy his machine, giving only his observations to the world. On the contrary he published a full description of all his apparatus and all his methods with diagrams, and plates, and tables of laboratory findings.

All scientists do this. Reports of scientific investigations consist very largely of descriptions of the machinery used, how it was used, and what happened when it was used. Theories and interpretations follow, but never stand alone. If they were published alone, no other scientist would credit them. One of the boasts of scientists is that in science nothing is ever taken on faith, since every research must be capable of repetition, and is not fully credited until it has been repeated by others than the initial investigator. Often reports of scientific research are not even published by the scientific journals until all the apparatus used (and described in full) has been duplicated by a referee with similar results. This boast overlooks the fact that the public, who can not repeat elaborate research, must and do take the whole business on faith. But it does emphasize as nothing else can do the importance of machinery in science. What scientists mean when they boast that all scientific research can be repeated is precisely this point, that the real achievement is the invention of the instrument. Any one can take the readings.

SCIENCE AS INVENTION

Thus the facts upon which science rests turn out to be machines. In the beginning is a machine—say, for example, the famous oil drop machine in which minute particles of oil of measurable size are sprayed into a vacuum and certain "rays" are allowed to enter. This machine is most artfully designed to permit microscopic observation of the movements of the drops of oil, and calibrated measurement of those movements, with readings in numbers of units. Such a machine, once designed, can be constructed by anybody. Any scientist anywhere can observe the oil drops, and doubtless many have. Here, then, are facts. But in addition to these facts there is a large and interesting body of theory—or law. According to this law, these rays have for some time been supposed to "free" electrons. Various things are supposed about the character of these electrons. Everything fits together very prettily. Previous suppositions give rise to certain "speculations" as to what electrons would do under such and such conditions, and upon one set of speculations of this highly theoretical character the oil drop machine was projected, designed, and brought to successful termination in the observed readings. This consummation is supposed by scientists to "prove" the theory.

Whether it does or not is neither here nor there. It seems to. Something peculiar has been happening in all sorts of apparatus of this type, some of it intended, some accidental. The electron theory is a most ingenious "explanation" of it all, and is cer-

SCIENCE: THE FALSE MESSIAH

tainly true to this extent, that reasoning upon it as a basis, various additional happenings have been provoked which thus far "fit" the theory. What is most interesting, however, is that the machines are paramount. The theory may or may not be "true." Electrons may exist, or they may be as mythical as the late lamented phlogiston, the supposed inflammable constituent of combustibles. The answer will be read in the machines. If the machines say no, no it is. In science, to be "true" means to be mechanically demonstrable. Science, we say, eliminates the personal equation. It does so by substituting the mechanical equation.

Thus the sum and substance of science appears to be that it begins in machinery and ends in machinery. This suggests that the function of scientific reasoning is to intermediate between one invention and another. Indeed, this conclusion is inescapable. Science as a body of law is quite incapable of propagating itself. Released from its obedience to machinery, it would soon soar into the region of pure imagination and so would speedily become—what theories about the elements and firmament have always been in other civilizations—just so much mythology, or magic, or folk-lore. Under the rule of the machine, the truths of science are compelled to make themselves useful—to the machines. They have accepted mechanical demonstration as their test; therefore they can never rest with theory, but have got always to produce new machinery to justify their own existence. As truths, or theories, they continue to be folk-lore. That is, they are still, in spite of their connections, a body of beliefs. But as a progressive body, they can never rest, but must always busy themselves with mediating between the machine that brought them forth and the machine that will give them proof.

I am alpha and omega, saith the machine. What the intermediate letters amount to in this scheme of things is another question.

CHAPTER IV

THE LURE OF MACHINERY

IF SCIENCE derives its potency from invention, invention in turn derives its power from the insidious penetrating force of all technical innovations. How civilizations change has not yet been reduced to the form of a universal scientific law. This is a genuine misfortune, but one which we can help to rectify by suggesting to the anthropologists the form which such a law might take. The difficulty seems to be that civilizations change as a result both of importations from other civilizations and also of internal developments. In recent years, the anthropologists have been so much interested in arguing which of these two processes is the more important, and particularly in proving that all civilization either was or was not exported from Egypt, that they have not thought to inquire what factor, if any, is the key to both the methods. When the time comes for its discovery, that factor will be found to be the technical culture trait. When two peoples come in contact with each other, they exchange most readily what is most readily exchangeable, namely physical objects: tools, weapons, foods, bric-a-brac, and the like. These things enter a civilization to which they are new. A place has to be made for them. Thus a series of adaptations is set going—and the two processes are completed. Only a philosopher would attempt to say now whether the ultimate effect was due to exchange or adaptation. Thus tradition has it that our three chief exports to uncivilized countries are Bibles, rifles and rum. This is enough to summarize the influence occidental culture has exerted. Meanwhile our chief imports are tea, silk, waiters, and laundrymen. Here we have the influence of the exchange upon ourselves.

This could be called a materialistic theory of history were it not for the fact that uses are always imported with materials. A musket suspended in space is not a culture trait. But muskets are seldom found suspended in space. They are more likely to be in use. Even if they are dug out of an old burial ground, they are not exactly in a vacuum. There is a skeleton alongside, and the musket, having obviously been buried with the corpse, still bears some relation to it. These Indians, we are bound to say, had taken over the use of muskets. That is one culture trait. They had also since long before practised certain rites of burial, always laying the weapons of their heroes alongside their bodies in the grave to be a protection to their spirits in the happy hunting grounds. Thus the buried musket is not only behaving properly as a firearm in use; its use illustrates at least one other culture trait beyond the actual fact of using firearms.

SCIENCE: THE FALSE MESSIAH

These are folkways, as no anthropologist would deny. In a sense they are nothing but folkways. The musket does not dictate how it should be used. It is only an inanimate object, and regarded as an inanimate object it is not even a culture trait, let alone a folkway. Civilization is a matter of behavior, which is to say, in this case, use; and the use is determined by the rules governing behavior, which is to say folkways. But it so happens that William Graham Sumner, when he put the word folkways into our vocabulary, made no distinction among the various kinds of folkways. He was concerned only to drive home the one outstanding fact that throughout civilization custom reigns supreme. The anthropologists, on the other hand, have been for the most part historians. Their analysis of ancient and modern civilizations has been historical. Consequently they have been compelled to distinguish between folkways.

Biologists have a theory that all life comes from life. This is like saying that all civilization comes from folkways. But while it implies a sort of evolution among the various forms of life, it does not afford much help in sorting those forms out and determining what came directly from what. For that, a study of the organs of the body is necessary, and a method of naming them so that species which possess a given organ can be set apart from those which do not. Culture traits serve this purpose in anthropology. They are the basis on which folkways can be named and classified, and their history traced from one civilization to another. A musket is only a physical object; but use-of-muskets is a culture trait, a little group of folkways associated with a central object, impossible without it, always passed from people to people along with a dominating object, and therefore, as seems only reasonable, named for it.

In some cases, of course, objects change hands only to be adapted to brand-new uses. Some Zulus steal a ladder from the trader's compound only to set it on an altar and make it the idol of a new religion. This could hardly be called a variant of the ladder culture trait. Neither can the use we make of imported totem poles. When the Romans conquered Greece, they stole all the statuary they could lay their hands on and took it to their villas in Herculaneum where they set it up in surroundings not entirely dissimilar to the Greek originals. There may have been important differences. Originally these statues may have been objects of worship of a sort, or markers of the graves of ancestors. That element is lost. But they were at the same time statues intended to be used in given surroundings—surroundings which the Romans carefully (or slavishly) maintained. When the nations of modern Europe plundered Greece, however, they made a quite different use of the objects of art which they removed "for preservation." The French took the statues and set them up in the Louvre. The British took the frieze of the Parthenon and set it up in the British Museum. Not by

SCIENCE: THE FALSE MESSIAH

the remotest stretch of imagination can we suppose such uses to have been intended by their originators. Phidias, the sculptor who designed the decorations of the Parthenon, may have been a free-thinker. Artists usually are. He may have been quite indifferent to the worship of Pallas Athena. But he was not indifferent to the fact that his frieze was planned as part of the entablature of a certain type of building. He certainly would be astonished to view it in its present setting, just as astonished as would be the priests and craftsmen who embalmed and decorated the corpse of Tutenkhamon at the motives and intentions of the men who are now busily rifling his grave. Their own motives are said to be open to question; all that gold and treasure may have been laid on in such profusion in part to prevent a certain wicked uncle from capturing it. But that it should finally come to rest in a museum in Cairo, with British Tommies standing guard at the door, was no part of the intentions of any of the mourners of King Tut.

Culture traits are always tricky. No two peoples ever made exactly the same use of any object. The more complicated the uses of the object are, the more chance there is for variation; and the more spiritual those uses, the more complicated they are likely to be. The employment of the wedding ring is more difficult to explain than the use of the teething ring, which is purely utilitarian, or even the ear-ring, which is decorative but not generally symbolic. The use of the cross in Christian countries is far more difficult, though rings may be as ancient as crosses, because the wedding ring applies only to the relation of husband and wife, whereas the cross concerns the religion that extends over the area of marriage and beyond it, covering ultimately all the institutions of society with a religious atmosphere of sacred creeds and observances; and whereever religion goes there is the cross going on before. Some say that the cross is a symbol of the actual historic crucifixion and nothing else. If it resembles the insignia of other peoples, that is an accident. It is a distinct culture trait of the Christian peoples. But others say that this is too simple; the cross in one form or another has been a sacred symbol from the cloudiest antiquity; crucifixion as a form of execution was itself a symbolic killing to which the cross brought its quota of significance. Meanwhile still others say that the history of the cross is much plainer than the history of Christian crucifixion; the reported execution of Christ upon a cross is probably a myth; and the myth was suggested in large part by the occult meanings of the cross. It was appropriate that the Messiah should die upon a cross; hence the myth that Jesus did actually die this occult death. So universal is the cross-swastika culture trait, and so intangible, obscure, and various, that the introduction of Christianity among a heathen people by missionaries is almost sure to be followed at least at first by the incorporation of the mysteries of the crucifixion and the Holy Virgin into the very midst of heathenism

SCIENCE: THE FALSE MESSIAH

where they flourish as new incarnations of ancient mysteries already practised by these heathens—mysteries so ancient that they have come down from a past which they and Chiistianity share, that past from which each drew its peculiar substance.

Nevertheless, some culture traits are perfectly distinct. Their movement to and fro can be traced with chronological precision. For example, there were no horses in North and South America before the coming of the Spaniards. Cortez and the others were received with fear and awe as a strange centaurian species—until they dismounted. Then it appeared that riding a horse is no very difficult feat. Any one can do it, and the Indians began forthwith. Horses were taken and used and bred. Some escaped. They were able to subsist in the wild state; they multiplied and spread, and formed the source of the great herds of wild horses which have inhabited this region ever since. Thus horses became a part of the culture of the American Indian. But it is interesting to observe that the particular uses to which the Spaniards put their horses, their trappings and accouterments, were transmitted along with the horses themselves. Thus it was not merely the horse as an animal, but the technique of horseback riding, which the Indians took over. They took it over rapidly and easily because horseback riding was very useful to them in the native occupations of their Indian existence.

Furthermore, none of them expected riding to alter their lives in any other particular. Savage peoples are just as reluctant to have their lives altered as any of the rest of us. In the exchange of rum, rifles, and the Bible for ivory and furs, the Bibles are usually thrown in. They are usually distributed by missionaries and not by traders at all; but they are given away. The reason for this is partly that the savages do not see offhand just what use they are going to make of Bibles. These books do not fill a long-felt want, at least so far as the feelings of the savages go. Moreover, the medicine men complain of untoward after-effects that may follow in their wake. In this they are assisted by the missionaries who claim quite openly that if the savages will only learn to read this book their whole lives will be changed. The missionaries are strong for the change; but the savages—being very much like us when socialism is in the wind—decline to consider such a thing until they are overtaken by actual calamity.

Rifles and rum are not subject to these discounts. They are so evidently useful that no one but a crank could object to receiving them. No strings of any kind are attached to them, so far as the purchaser can see. We may be able to see farther, to observe that after the savage has acquired a taste for rum he is obliged to barter all his furs for more rum, and when they are gone to put the rifle to use to obtain more, with which to get more rum. Probably the bargains are made more or less in the aroma of alcohol, so that the ratio of furs to rum and ammunition is always somewhat unfavorable to furs. As

SCIENCE: THE FALSE MESSIAH

the result of this process, the savages gradually find themselves completely at the mercy of the traders. Their tribal life revolves about fur-trading as its center. They have ceased to be an independent civilization and have become an outpost of European culture. But this is happily veiled from the first savages to drink rum and hunt with rifles. So far as they can see, the future is to be precisely like the past, except that the superiority of rifles over spears will make it more so.

This is why technical innovations always come easy. Whatever rum is—and we may as well put it in a special class as a habit-forming vice—saddle-horses and hunting rifles are technical appliances. In use, each is a technical culture trait. Such traits pass very readily from one civilization to another. Perhaps it is too much to say that all technical culture traits outdistance all other aspects of civilization in fluidity, especially since an exchange of habit-forming vices seems to be the earliest and most contagious effect of contact between two civilizations. Neither do all technical traits pass from hand to hand with equal ease. Cameras make little headway in Mohammedan countries, where all pictorial reproductions are forbidden by the most sacred *mores*. Bicycles are not popular in Tibet, where all roads run up-hill. Automobiles can be used only with difficulty in lands where there are no roads, no garages, and no filling stations. But barring these special disabilities, it still remains true that technical appliances can be introduced where nothing else will penetrate, and that people will learn to use them who would scorn to imitate the outlander in any other way. The rifle, as aforesaid, is now in use all over the world. Savages of every breed seem to take to it intuitively, recognizing in it the savage culture trait par excellence. The English have found it easier to build several thousand miles of railway in India and to make travel by rail a commonplace for Hindus of all castes, than it would have been to enforce a single item of the program of the missionaries—even such harmless items as the British notion of decency of dress, or the European idea of bodily cleanliness. The railway is unpleasant. It scars the landscape. But it invades neither temple, nor home, nor caste, and by its agency famines are checked and families united.

This is the universal history of technical innovations. Such things are useful, and they appear to be harmless. Whatever effects they have of a general nature, touching the civilization as a whole, are surreptitious effects, stealing upon the community quietly and seldom striking the men who actually introduce the culture trait. Our printing and gun powder are Chinese; but we received them without taint of that pacifism with which the Chinese are afflicted. Our mathematics is Moslem; but we drove out the Moors, Saracens, and Turks, bag and baggage without taint of that polygamy with which Mohammedans are afflicted. Our industrial revolution began, as some historians say, with half a dozen technical improvements in the textile industry; and

SCIENCE: THE FALSE MESSIAH

it took us a century to realize that anything of moment had happened to us, beyond the obvious improvement of spinning and weaving. Most of us still imagine that we shall continue to worship at shrines that were holy before the day of the power loom. Some of us think not. But none of us hesitates to install a radio, or to exchange his used car for a later model. Why should we? Are not these things obvious improvements? Such is the lure of the technical culture trait!

CHAPTER V

A KINGDOM OF MACHINES

AFTER Samuel Butler, in his character of adventurer in *Erewhon,* had passed over the range and descended to the habitations of that interesting but paradoxical people, the Erewhonians, he was promptly arrested and thrown into jail. The crime with which he was charged was that of carrying a watch. As he found out later, the Erewhonians had had their fling with machinery. They had passed through their industrial revolution and emerged into a machine age. Then they had found themselves becoming completely enslaved to the machines with a fair prospect of being reduced to the position of Yahoos, with machines instead of horses as their task-masters. But before it was quite too late they had revolted, thrown off the yoke, destroyed their oppressors, gear, piston, and dynamo, wiped out all traces of their servitude, and—to prevent a recurrence of the nightmare—made the manufacture and even the possession of any machine a high crime and offense against public morals, punishable by imprisonment and death.

This interesting transition and return to normalcy was revealed to him much later in his stay

SCIENCE: THE FALSE MESSIAH

in Erewhon, chiefly through the work of an ancient Erewhonian philosopher known as the *Book of the Machines*. It seemed that machinery had stolen upon the Erewhonians gradually and insidiously, at first after the manner of technical culture traits throughout the world. One invention was borrowed from a neighboring people on the right. Another was taken over from wandering traders from the left. Still another was brought in by shipwrecked sailors stranded on the coast. Each was highly ingenious, and provoked the wonder and admiration of all who beheld it. We can imagine these as the trick of printing by pressing paper on an inked block; the trick of extracting glass from sand and using it as a transparent window covering; and the trick of putting saltpeter and charcoal together to make a noisy and showy combustible stuff.

The Erewhonians were an intelligent and ingenious people. They looked upon these ingenuities with awe and wonder; but they began at once to improve upon them. Printing had been imported, let us say, from the Chinese, whose written language is constructed without benefit of letters, so that each printing block required to be carved afresh. But Erewhonian, like English and the other European languages, is written by the use of twenty-six interchangeable symbols. These twenty-six letters, repeated over and over in ever varying combinations, were sufficient to render all Erewhonian literature into writing. Hence the first improvement of movable type was obvious almost as soon as the Chinese trick of printing was introduced. Similarly, the explosive mixture of charcoal and saltpeter was soon put to a nobler use than its originators had ever dreamed of in the art of war. It seemed that the Erewhonians already possessed a mechanical appliance which they had developed from the bow and arrow: an engine for throwing arrows much farther than a hand-bent bow could send them, called an arbalest. This at once suggested a similar tube for throwing missiles by use of the propelling power of the new explosive, which was soon developed and called the arquebus. As to glass: no sooner was its manufacture under way than the workmen noticed that drops of glass in the shape of a pea, or lentil, refracted light, and that by shaping the lentils of glass into lenses the refractions could be controlled. From this the way was open to the technique of magnification.

As we can imagine, these developments led to many things. Printing from movable type made reading matter so common that reading and writing became general, and social classes were dislocated. So effective was the arquebus that the art of war became a larger undertaking than it had ever been before; larger military units were required to handle the more effective weapons. As a result, the Erewhonians became a highly centralized "modern" nation. Meanwhile lenses made the way for science.

None of these results had been expected in the least when the new contrivances were introduced, and it seemed highly doubtful if the Erewhonians,

SCIENCE: THE FALSE MESSIAH

who were a very conservative people, though perhaps not more so than their neighbors, would ever have tolerated them if they could have foreseen the uses to which they would shortly be put. But more disconcerting still was another characteristic of the new inventions. It appeared later that they had the power of fertilizing each other. This was not noticed at first. When the combination of the fire-cracker explosive and the arrow-throwing arbalest first occurred, the men who had witnessed the new development took credit to themselves for the achievement, recorded their exclusive authorship, and boasted openly of their superiority to the Chinese in mechanical inventiveness. But after machines had multiplied in the land, people began to notice that wherever complicated mechanical contrivances were brought together new and more complicated contrivances were the invariable result. Each new machine suggested and immediately brought about a dozen improvements, and each improvement soon found a dozen applications besides its original one; so that new hybridized varieties of machines commenced to spring up on every hand.

One important consequence of this mechanical evolution was the tremendous increase of the power of reproduction among the machines. The earlier and simpler machines spawned by budding. Each individual machine served as the pattern for the creation of another like it. But later on, when larger and more complicated machines had arisen,

A KINGDOM OF MACHINES

certain types of machines were specialized to reproduction. These settled in machine-shops, as the lying-in hospitals were known, and brought forth each according to its kind but at a rate of reproduction far in excess of any animal. This, indeed, was the strange paradox of mechano-genesis, as the reproduction of machines was called: whereas among animals only the smaller single-celled bacteria reproduced anything like instantaneously, the larger animals requiring longer periods of gestation in proportion to their size; among the machines, on the contrary, the larger the machine the more rapidly it seemed able to reproduce itself. One motile species, for example, known as quadrurotes, increased to more than twenty millions within a quarter-century of its first appearance. A single breed of quadrurotes, called in Erewhonian the Ledom T Drof, produced something like eight millions in less than twenty years. At the height of their fertility, the machine-shops of the Ledom T Drof spawned at the rate of a birth a minute, or one hundred thousand a month. As might be supposed, such a birth-rate alarmed the more foresighted Erewhonian economists, and statistical studies were brought out which revealed a condition fraught with danger to the entire race of Erewhonians. One scholar, a gentleman named Suhtlam, computed the increase of machines by mechano-genesis to be taking place in geometrical proportion, the machines doubling in number every five years. Meanwhile the native population was doubling only by twenty-

SCIENCE: THE FALSE MESSIAH

five year periods. The ominous conclusion of this research was that there would soon be too few Erewhonians to provide sustenance for the machines, so that all would go down together, the parasites perishing with their hosts.

No doubt the revolution which ensued, in which the machines were totally destroyed, was brought on by contemplation of this awful prospect. But the precise stages by which it came about were left obscure in the *Book of the Machines,* and Samuel Butler was never able to repair this important gap. This is particularly unfortunate since the history of our own times as yet affords no parallel. Among us, machines are not only still in the highest honor; they seem likely always to remain so. The reason for this is that each separate machine seems to be quite harmless and extremely useful. This is what leads to the introduction of the first elementary machines as transplanted culture traits borrowed from another people. The novelty is an inanimate object which yields readily to human control. There is nothing about it to suggest danger, so long as it is properly handled, and we never doubt our ability to handle it properly. Moreover, we can see at once various uses to which it could be put, and the advantages that will accrue to us as its introducers if we are the first to employ it regularly. If the innovation were a new style of parting the hair, or a new method of domestic discipline, the case would be quite otherwise. Any one who ventured to take it up would be branded as a dangerous radical, and

A KINGDOM OF MACHINES

stoned and ridiculed out of existence. But the first man to use a new machine is hailed everywhere as a Progressive Citizen. Probably this has always been so and will always be so everywhere except in Erewhon.

This psychology—the state of mind in which simple peoples borrow inventions from their neighbors—is not affected in the least by the increasing prevalence of machines, any more than the attitude of parents toward their children is altered by the increase of population. Statesmen orate about race suicide and the duty of parents to the state; but parents have children because they want them, not because they think the state will like it. And by the same token parents love their children just as much in thickly populated sections as they do on the fringe of civilization. Hence the difficulty of reducing the population. Each individual, and all his friends and relations for him, resists a reduction which begins with him, and does his best to stay alive in defiance of the greatest good of the greatest number; and if it were not for natural decay and dissolution all the descendants of Adam would be alive at this day—with the possible exception of the sorrowful Werther. This is why Malthus hailed war and pestilence as the saviors of the race. The increase of machines is regulated by the same principles. No out and out reduction is even thought of because no single machine could possibly be spared. In addition to the inviolability of mechanical inventions, which operates like the inviolability of human

SCIENCE: THE FALSE MESSIAH

life in behalf even of the least worthy, each machine is some one's property, just as each individual is some one's son or daughter, and could be put out of existence only by confiscation in the face of the bitter opposition of the owner and all his class. Only rust and traffic accidents keep the machine population in check.

Among human beings, there is a negative check called neo-Malthusianism: control of population by checking the birth-rate. This is theoretically possible among machines. For example, by the refusal of patents, building permits, and the like, a government could (theoretically) prevent any addition to the present technology. But in actual practise such a step would no more be possible than it would be for a government to prescribe, or even openly allow, the general use of contraceptives. The difficulty in each case is psychological. The sacredness of human life is supposed by all proper people to extend even to the unborn; so that to discuss the subject of birth control is indecent; to practise it is immoral; and to extend or in any way facilitate its use is illegal and criminal and punishable by heavy fine and imprisonment. Fines and imprisonment have not been imposed as yet on those who advocate a discontinuance of invention. But perhaps the reason for the oversight is that almost no offenders have appeared. No one has even thought of putting a check on mechano-genesis. The very idea is preposterous.

Indeed, the repugnance to neo-Malthusianism

A KINGDOM OF MACHINES

goes deeper, and touches on the confines of religion. Not only the sacredness of life but the sacrament of marriage is involved, and that is a matter of deep concern to the church which ordains and guarantees this sanctity. Under the canons of the church it is the duty of man to "replenish" the earth. Any hanging back from the responsibilities of parenthood is to be interpreted as an affront to the church, to the most sacred injunctions of the Almighty, and to the Omnipotent Creator Himself. At the very least it is "tampering with nature," and is not to be thought of.

The sacrament of technology is science. From the very first, the adaptation of machinery has required certain calculations and predictions, a theory of "what is happening" from which we can infer what is likely to happen as a result of modifications in this or that machine. As the machines have become more complicated, the calculations have become correspondingly complicated and extensive, the theory of "what is happening" has taken in more and more events until it has come to be a matter of very grave importance. It has become sacred as The Truth. Just as the sacredness of human life extends not only to every human being but even to the potential human beings contained in the germinal cells, so the sacredness of Truth extends not only to the words of children to parents, the words of witnesses before the bar of justice, the words of prophets to their flocks, the words of God to man, but even to the words of calculators to their mechanics. Indeed, strange as it may seem, the latter have come to be considered, in this age of machinery, the most sacred of all. Heresies against the church are a flying in the face of God; heresies against science are a flying in the face of nature. The calculations of science are the revelations of the prophets of nature, and any attempt to check or curtail them—such as would be necessary if invention were to be curbed—would be a tampering with nature intolerable to every pious soul. No relief from any possible surfeit of machines is to be sought in this direction.

Of course we may never experience such a surfeit. We may be able not only to support more machines than the Erewhonians but to control them as they could not. The probable cause of the Erewhonian revolution, hinted by Samuel Butler in his famous book, was the sense of degradation which the men of Erewhon finally came to feel at their complete dependence on machinery and domination by it. How they could throw off a technology on which they were so dependent is still a mystery. We probably could not. Even at the degree of dependence which we have now reached we should be quite unable to support our own population of human souls if we were to wreck our whole technology in a burst of revolutionary temper, and as time goes on we shall be less and less in a position to chance it. Nevertheless we may escape being dominated. Perhaps the whole idea is just a figure of speech invented by a clever literary man.

CHAPTER VI

INDUSTRIAL REVOLUTION

"THE industrial revolution" is a tricky phrase. Its pronunciation is obvious; but at least one eminent economist, Walton Hamilton, insists that it was not a revolution, it was not industrial, nor was there only "the" one. Aside from these reservations, he says, the phrase is perfectly true and just.

Nevertheless, even allowing for these lapses from strict accuracy, the expression "industrial revolution" is decidedly valuable. It was invented, or at least brought into general currency, by Arnold Toynbee, about half a century ago. He was in revolt, as young economists always are, against the iron-clad principles to which the opinions of the classical writers had been reduced by much repetition, and he wanted to refer his hearers instead to the common facts of economic life, such as any one of them might see and study simply by paying attention to what was going on around them. He also wanted to lay especial emphasis upon the growth of manufacturing, and the transformation of England from a rural countryside into an urban, mining, smelting, and manufacturing establishment. This transformation had occurred first in a group of related, key industries. By a series of closely related and almost simultaneous inventions, the textile industry—already the key industry of the United Kingdom—was suddenly transformed from a handicraft to a machine process. Power looms and spinning jennies were introduced, steam engines and power transmission were developed for the first time upon what we should recognize as a commercial basis. Coal was mined and ore was smelted in commercial quantities and a beginning was made in the manufacture of machinery. With this sudden rush, the industrial era was begun.

The transformation of the textile industry was highly dramatic in two different though related ways. From the technical point of view, the changes were sudden beyond precedent. No such transformation had ever occurred before. The invention of printing, for example, or of gun powder, may have been momentous. Each doubtless brought vast social changes in its wake. But great as is the contrast between a scribbling monk and an ink-begrimed type-setting craftsman, that contrast is not so suggestive of miracles and magic as the difference between the spinning wheel and the hand loom, such as Silas Marner carried about with him from household to household, and the coal-fired steam boiler, the sliding cylinder and piston, the revolving belts, and the prodigious clattering speed of the power looms and jennies. Beside these awful realities, the explanation that printing was still a handicraft no less than pen-and-ink copying,

INDUSTRIAL REVOLUTION

whereas power weaving was a machine process, seems weak and ineffective. These are the right words. But words are pale reflections of reality. The effects of printing may have been greater; but the appearance of a factory is more spectacular.

So is the appearance of a mill town. We have no way of measuring the subtle influence of type or gun powder upon civilization, nor of power looms. Gradual and insidious effects may be more important in the end than those which can be seen at a glance; but they will not be so exciting. The social effects of the transformation of the textile industry were exciting. They took the form of a sudden dislocation of the population: mushroom towns sprang up at the mining and manufacturing centers, with the confusion and hardship, squalor and universal grime to which we have since become accustomed as the inevitable by-products of sudden growth. All these things were very obvious, and to many people very annoying. Like the miracle of machinery, they gave dramatic quality to the changes enacted in the textile and related industries.

But the first act of a play is not necessarily the most important one. Indeed, the unreality of stage plays is due partly to the fact that they can represent only the thoughts and feelings or at most the culminating decisions of characters who must be supposed to have been alive for years before the action of the play begins, and to have acquired the various prides and prejudices, out of which the play has come, through a long succession of earlier

SCIENCE: THE FALSE MESSIAH

years. In imagination, at least, the action of a play extends over a long period which may or may not be adequately recognized in the final action. In real life this must always be the case. A play may fail to motivate the actions of its characters so as to allow for their previous existence. It is then an artificial play. Actuality is never artificial. Whether we understand it or not, preceding conditions—and preceding conditions alone—have made possible the current events which we are witnessing. If these events startle us, that is only because we have neglected to notice what has long been going on before our faces. In this sense, the distinction between evolution and revolution is merely between what we have noticed and what we have failed to notice. A revolution is the dramatic climax of a long and gradual process which we have overlooked.

The industrial revolution was no exception to this rule. Clearly, the sudden transformation of spinning and weaving could have come about only as a result of a long train of inventions—a train so long that a number of simultaneous applications could be made. Coal mining, iron smelting, power generation, the spinning and weaving of wool are very different operations. It is incredible that each should have been solved by quite separate acts of genius. Ordinary sense requires another explanation; the progress of invention in general, over a long period, had reached a point at which a series of applications could be made. Just these applica-

INDUSTRIAL REVOLUTION

tions were made because of some other condition which made such a combination of applications simultaneously necessary.

That condition, again, refers back to another train of past events. The production of woolen goods is a historic industry in Great Britain. Its development is a story extending over some hundreds of years, a story much too long and complicated for telling here. Presumably the industry was under certain pressure, and it passed on a part of this pressure to the inventors. Anyhow, without knowing anything more of the situation than that the woolen business was already flourishing while the camera business had not yet come into existence, it is easy to see that the application of machine methods to woolen manufacture rather than to kodaks was a natural and obvious development of earlier conditions. Thus the spectacle which Arnold Toynbee called an industrial revolution must be understood as a dramatic moment in the lives of two very ancient processes, invention and woolen manufacture.

But if the beginning of a play is an artificiality, the end is downright prevarication. In the drama, the evil consequences of the villain's acts or the hero's weaknesses, are supposed to stop work with the death of all the principal characters. Such an air of finality does this simple device lend to any plot that it has been employed by all the greatest dramatists from Euripedes to O'Neil. Yet it is pure nature-fake. In reality, no final curtain ever

SCIENCE: THE FALSE MESSIAH

interposes its asbestos barrier between hell fire and the steady march of the victims of human villainy. There is no last laugh.

This fact is even more obvious than the fact that every present has a future. Every one knows that every present, however complete and satisfying it may be, will inevitably have a future in which it may appear quite different. Yet it is difficult for any of us to realize all that this knowledge can imply—simply because though we have lived through the past, none of us has yet lived in the future. To every one the present is the limit of experience. To every one, therefore, the present is the standard of measurement, the test to which all things are subjected.

From the point of view of the present, the dramatic change which overtook British textiles in the last quarter of the eighteenth century, does mark a pivotal moment in modern history. Since that time we have more or less fully realized the application of machine methods to all branches of manufacture. This is the Iron Age. Coal is King. The factory is the modern temple. Consequently, the moment when these forces first asserted their ascendency is the breaking point between the old order and the new.

But in making that assumption we assume not only that Coal is King, but that his reign is going to be long and undisputed. This need not necessarily be the case at all. Let us introduce Giant Power into the discussion. Suppose for the moment

INDUSTRIAL REVOLUTION

that the chief source of power—perhaps ultimately the only source of power that will be able to compete—is shortly going to be water power, and later wind power and tidal power, and that these sources of energy are going to be tapped by electric generators, organized into a universal system of power transmission and storage. Let us suppose further that iron, which has already been transformed by later metallurgy into a multitude of alloys far superior to pure ore, will shortly be displaced by a still more common metal, aluminum, rendered hard and tough by similar alloying. What of the "industrial revolution" then? Coal will be a worthless smut. The steam engine will be preserved only in museums, like the exhibit on the arcade of the Grand Central Terminal of the first cars and locomotive of the New York Central Railroad. Iron machinery will also be on the scrap heap. Wool will have been retired from use by the synthetic fibers of which "rayon" is a forerunner. From the perspective of that "present," the early experiments in spinning and weaving wool by machinery will appear to be abortive and mistaken inventions, altogether inferior and unpleasant, an unsuccessful experiment which was later fortunately followed by a more genuine and effective "revolution": the giant-power-electricity-aluminum revolution occurring about the middle of the twentieth century.

If we suppose further the disappearance of factories and of railroads and possibly all crawling transportation, the artificiality of our present point

SCIENCE: THE FALSE MESSIAH

of view becomes far more striking. And these suppositions are not chimerical by any means. Giant power is already a matter of solicitude to legislatures and, we may guess, to great electrical corporations. Electrical power differs from coal as a source of energy in this: it can be used in power units of any size with equal efficiency. A motor-driven sewing machine is just as efficient as an electric locomotive. Consequently it is not necessary for the utilization of electrical energy that machines be assembled in a single plant of vast dimensions. So far as the power is concerned, each machine might as well be in a separate building—for instance, in a work room connected with the operator's home. Thus giant power may effect a demobilization of industry once more from the factory to the home, and from the city to the countryside. The great bridges, railways, etc., which no less than great buildings have required tremendous steel mills for their manufacture, may all be retired from use by air transportation. Thus the demobilization may actually be complete. If all this should occur, it would then appear that even the building of cities and the development of urban life (the characteristic feature of recent civilization) was an impermanent phase and not the permanent condition of life under the machine technology; so that whatever social adaptations we have made to machinery thus far would turn out to be as abortive as our steam engines. The full development of the mechanical arts is in the future, and the civilization which will be required by the mechanical way of life is equally in the future. Thus it appears very possible indeed that the industrial revolution, instead of having transpired and worked itself out in the seventeen seventies, has actually not yet really begun. We have seen only its preliminary stage, the first act of the drama.

There is still another reason, and that the most important one of all, for believing that we have witnessed only the opening of the play. Whether our industrial system has reached its final stage or not, whether the future of machinery is to be the same or much different from what we already know, nothing is clearer than this: even as it is, it has not yet fully had its way with us by any means. On the surface modern life is very different from what the human race has ever known before. These tremendous cities and thundering machines, these specialized occupations and standardized enjoyments, give an appearance to life very different from any it has ever had before. Nineveh and Tyre, Athens, Carthage, Rome, Byzantium, Paris of the time of Abelard, Florence of the time of Leonardo, London of the time of Shakespeare, were all very much of a piece compared with New York of the time of Edison. The differences among them are great; but they are not so great as the difference between each one and New York. Not even the differences between any of these and the aboriginal cities of Egyptian, Minoan, or, for that matter, Mayan civilization compares with the difference be-

SCIENCE: THE FALSE MESSIAH

tween Shakespeare's London and New York. The industrial revolution represents a major cleavage in the history of civilization, a fault line not between two hillocks but between two mountain ranges. What lies beyond is somehow separated from all that lies before.

We do not know what lies beyond, because in our journey through history we are still carrying the baggage we have brought with us from the nether regions. We are not yet really living the life of modern civilization. We are living as Europeans have lived in the past with slight allowance for such modifications as have been forced upon us by physical changes that could not be ignored. These modifications appear to be many—from the point of view of a present in which we still think of the industrial revolution as the invention of the spinning jenny. But the most casual survey of our living arrangements will show that though we go to work on the subway, and perform labors which no Elizabethan could have comprehended, we come home to a fireside not greatly unlike Anne Hathaway's, eat a dinner which she would have relished, since for all our tropical fruits and Burbank vegetables, its base and substance is still the beef and bread of Merrie England, and spend the evening playing a game but recently called "whisk." Our machinery is modern; but our institutions are medieval. They are changing—into what, we do not know. For the present we think of the process as the "deplorable loosening of modern life."

INDUSTRIAL REVOLUTION

Manners and morals, we say, are being "relaxed." We do not reflect when we say this that those manners and morals belong to an era which we snub with the epithet "dark ages."

Now it is not reasonable to suppose that the institutions of the dark ages will consort through an indefinite future with the technology of the machine ages. Those institutions were developed through experience that was long out of all proportion to the brevity of our experience of machinery. Presumably it was gradually and finally adapted to the conditions which prevailed then, conditions not essentially dissimilar to those that have prevailed in Europe since the stone age. The institutions are not essentially different either. A new technology has arisen. To expect a new civilization with a new scheme of institutions, manners, and morals, adapted to it as the old scheme was adapted to the old technology, is only reasonable. What is unreasonable is to suppose that the commuter's train will continue indefinitely holding together two different and opposite civilizations on the opposite ends of the line: on one end, the occupations of the machine age; on the other, the institutions of the middle age.

Our present civilization is a hybrid. The middle age was quickened by machinery and the present hybrid is its fruit. Like most hybrids, it shows every sign of being **unfertile and impermanent.**

CHAPTER VII

THE NEW FREEDOM

THAT past from which the industrial revolution is so obviously loosing us is for the most part unwept, unhonored, and unsung. We do not like the past, or its dead hand upon the present, and the farther we get from it the less we like it—quite naturally. We call our dissociated past the "middle ages," and we make "medieval" a term of violent abuse. Our peculiar joy is novelty, and our delight is to let freedom ring. For better or for worse the new freedom is one of the chief constituents of contemporary life. It shows itself both in statecraft and in stockings.

There are various reasons for this, some more important than others. For one thing, the talk of the hour is always produced by the talkers of the hour; and the most fluent talkers are naturally the professional speakers and writers. Such people form a class by themselves. They are often said to be temperamental, by which they mean that they are not of ordinary clay and can not be expected to trudge along in the ordinary ruts, while we mean that they have an exaggerated opinion of themselves and consider that what is good enough for us is not good enough for them. Consequently the advanced thinkers of all ages have been abnormally interested in freedom. Probably there has always been enough oppression in the world to justify them at the Last Judgment. But they have always been more sensitive about it than any one else; and while this is not necessarily to be counted against them, it still must be remembered to give the Devil his due and put us on our guard against the rhetoric of professional exhorters.

Still, it must be admitted that very few apostles of freedom have been temperamental enough to rebel against all order whatever. Most of them have denounced some particular oppression that has happened to be rife, and have led the vanguard of revolt against it. Their idea has not been to wander on, aimlessly, year after year in the wilderness of anarchy; their leadership has been directed to some specific land of milk and honey where, the particular annoyance having been removed, human life will be richer and finer, more beautiful, and fruitful, and everything. This means, of course, that freedom has had a new and special meaning every time it has appeared.

The continual reappearance of the battle cry of freedom down through all the ages is therefore liable to produce a false impression, the impression that our past has been a steady succession of victories over tyranny; whereas, in fact, each conversion may have been followed by a relapse, possibly in some other direction. This may or may not have been the case. No rule can be laid down for either the affirmative or the negative. Each case ought to be treated on its merits. But the point is that a dogma to the effect that every conversion was followed by a relapse, and so that humanity has made no real progress toward freedom, is just as reasonable and just as unreasonable as the opposite dogma, that the frequency of crusades proves how much progress has been made. The frequency of crusades proves nothing at all except the frequency of crusades, each of which is probably a law unto itself.

These prejudices and preconceptions are especially hard to disentangle where our most cherished past is under consideration. For example, the English-speaking peoples have long indulged in a myth that the foundation of "Anglo-Saxon freedom" was laid at Runnymede, where King John signed the Magna Charta, in 1215. But the freedom that was sealed at Runnymede was a special kind of freedom. As many later historians have pointed out, the importance of this event was "discovered" by a British historian with a considerable gift for dramatizing history. What actually happened at Runnymede was a transfer of power from a single king to a group of feudal barons, a transaction which may have been good for the barons but almost certainly was of little or no use to us. In 1215 feudalism was in its heydey, and Magna Charta was far from being the signal for its collapse, or for the "forward march!" toward democracy—which was the last thing a Norman baron would have dreamed of, if any Norman baron was well enough read in Greek and Roman literature ever to have heard of such a thing. That blot upon English history, the War of the Roses, was far more serviceable to democracy (if we can trust these later historians), inasmuch as it killed off most of the descendants of King John's precious barons and gave "the people" their chance.

A similar high spot in the panorama of freedom is the "bloodless revolution" of 1688, when the "worthless" king was *spurlos versenkt,* and the highly moral William and Mary were installed upon the British throne by act of Parliament. This was a triumph both of freedom and tolerance—since James' undesirability consisted largely in the fact that he was a Catholic. In other words, "tolerance" in 1688 meant giving Protestants a chance, even at the cost of dethroning Catholics. "Freedom" similarly meant reducing the authority of the crown and increasing that of Parliament—very nearly the same Parliament which had repealed Cromwell's equalizing laws only twenty-eight years before. The expulsion of James II was a very special breed of liberation.

The new freedom is a mongrel cur in comparison. Certainly it is not the "freedom to worship God" of the Pilgrim Fathers. Nor is it the independence of the United States, upon which rests our official title as a free nation, independent of His Britannic Majesty. The modern declaration of independence

SCIENCE: THE FALSE MESSIAH

is more extensive and various than any of these. No single crusade is under way. Children are declaring independence of their parents; parents are declaring independence of their children; wives are declaring independence of their husbands, (and perhaps even vice versa). Laborers are demanding to be freed from their bondage to employers. Employers are demanding freedom from the restrictions of government. Authors and producers are demanding to be free of all restrictions, and teachers are demanding their "right" to run their schools and colleges to suit themselves and are politely requesting their trustees to walk the plank. The tyranny of convention is everywhere denounced. In sum, we are declaring our independence not of our sovereign but of our past. This is a new freedom.

Even so, the new freedom is distinguished only by its variety and magnitude. Every revolt is a revolt against the past. But as a rule it is some one feature of the past which pinches. That is not the case to-day. We have a labor problem. But we also have a divorce problem, a religious crisis, an educational revolution, and a "futurist movement" in all the arts. Each is just as acute as the industrial unrest and just as important for the department of life concerned, though all may not be so crucial for the immediate welfare of the community as a severe strike.

Many peaceably inclined writers have been soothing us recently with documents calculated to show that unrest is to be found in all periods. They

THE NEW FREEDOM

quote some prodigiously old Babylonian hieroglyph to the effect that everything is changing; things are not what they used to be, or Saint Augustine's caustic denunciations of the looseness prevailing in his time. These are excellent sedatives and no doubt they are wisely administered. No one is going to profit by becoming hysterical over our state of flux. Nevertheless, the admitted facts—that other periods have experienced transition, and that some people in every community deplore the passage of better and older days—need not be taken as a complete disproof of the evidences of transition which still exist before our eyes, whether other people have seen similar things or not.

Neither is the theory of evolution entirely satisfactory as a comfort in time of need. Huxley defined evolution in words calculated to satisfy scientists and confuse other people, the net effect being that it is a development from the simple to the complex. According to the formula, civilization also might be expected to become constantly more complex. So far so good: ours bears out the formula. But neither in biology nor elsewhere in nature is evolution a process of accelerating change, ending in continual flux. Higher animals are more complex than lower ones; but they are just as stable, so far as we can see. Under the evolutionary rule, therefore, civilization might become more complex. But there is nothing in the formula to justify its becoming at the same time more fluid. Sir Henry Maine, the great English historian of law, laid down the rule that the history of European institutions is a movement "from status to contract." But he did not intend to convey the impression that the change has been from status which can not be altered to contracts which can and will be broken. To a jurist, contracts are sacred. The development was from involuntary to voluntary relationships, both equally rigid.

The status-to-contract formula is an interesting one because it suggests what lies behind the new freedom. To go no farther back, the basis of European civilization is that scheme of institutions which we call medieval, or feudal. This can be pictured as a series of established relationships of a very personal character. The population was stable and, in any given locality, homogeneous; and each man occupied the station he did and pursued the vocation he did pursue by virtue of being that person. Status was attached to the man, and passed on to his children just as names are passed on to-day. Indeed, the naming system is a vestige of that order, a highly significant vestige, and the Lucy Stone League of women who refuse to abide by it is a very pretty straw in the modern wind. However, the change in which Sir Henry Maine was chiefly interested was economic in the sense that the determination of status ceased to be entirely (and at a later date even largely) a matter of being a certain person, and came to be a matter of owning property, or assuming a certain profession. No doubt this is a considerable change. Certainly the historic causes of it are extremely complicated. But one can hardly help observing that it does not greatly alter the status of anybody. Whereas a nobleman once occupied a certain relation to a sovereign, from whom incidentally be held certain estates as a sign and seal of that personal relationship, later he occupies the position of landlord (of those same estates), and by virtue of being a landlord is incidentally placed in a certain relation to the sovereign (in matters of taxation and so on). Whereas the men who worked his estates were once serfs bound to the land by their personal relation to him, the lord, later on they are tenant farmers and farm laborers, designated as such not by descent but by inheritance—or the lack of it.

So far as the nobleman-employer and the serf-tenant are concerned, this change has often been effected very rapidly with scarcely perceptible alterations of the actual lives of either party. In a Latin-American country, let us say, peonage exists such that the males of a certain class are bound to the soil and obliged to work it under the direction of a free-holder. Peonage is then suddenly abolished by law. The peons must thereafter be paid wages, or goods in lieu of wages. But having no other occupation they naturally continue to work as before, while not being any more skilful or provident than before they go rapidly into debt to their employer, who graciously allows them to do so, and to spend the balance of their lives legally wage-earners, but actually peons.

SCIENCE: THE FALSE MESSIAH

The economic "revolution," by which the shift of base from status to contract took place, considerably preceded the industrial revolution, and need not necessarily have been followed by it at all. Had it not been, European society would no doubt have gone on much as it had gone on in the past. The powers of wealth arranged the population very much as it had been arranged before. Wealth is no more diaphanous or volatile than patents of nobility. Poverty is no less ruthless a taskmaster than a bailiff. Apprenticeship as a cobbler or a lawyer fixes a man just as securely in one century as in another. But more important than this: no further changes need ever have occurred, barring further accidents. There is nothing about the capitalistic system to discommode the established life, for example, or to throw religion into turmoil. Capitalism is merely a codicil of feudalism.

Once this'is understood, it immediately becomes clear not only why the new freedom is more than a polite continuation of the orderly process of change that has been going on since Magna Charta, if not since Babylon or Eden, but even what the protest is all about. The protest is directed against the whole medieval-feudal-capitalistic system of personal relations and social institutions. In this it is as different from the "freedom" and "tolerance" of the bloodless revolution of 1688 as modern New York is different from restoration London. No doubt the historic process which lies between these two is a continuous one; but that does not mean that nothing else besides that continuity has occurred. What has occurred is—New York, and all that it stands for. This had not occurred in 1688. It had not occurred effectively when Sir Henry Maine surveyed the process half a century and more ago. The spinning jenny had occurred; but not having seen Times Square at half past five, Sir Henry Maine underestimated the importance of machinery. Surveying the scene from the perspective of his time and limitations as a jurist, he looked upon ownership and saw that it was feudal. He therefore recorded the change quite accurately as a movement from status to contract; and being an optimist he felt that this was progress.

The formula will not do to-day because the world of to-day is not only capitalistic but industrial. Having been at it at least since the time of Henry VIII, the Anglo-Saxon world is pretty effectively adapted to ownership as the basis of its social institutions, so that the family life, the spiritual life, the artistic life, even—in the main—the intellectual life of the old order have been brought forward with no more alteration than four hundred years effect at any time. At the end of this period, however, another and a far greater change had occurred. Its magnitude can be indicated quite simply by one fact: population. One hundred and fifty years ago (before the spinning jenny) the population of Europe was what a Cæsar might have dreamed. To-day, only the industrialism that has brought it forth could continue to support it.

SCIENCE: THE FALSE MESSIAH

This population finds itself living, and building machinery, under a scheme of institutions that was worked out through hundreds of years of stable (if not peaceful) agriculture, with the merest smattering of commerce and manufacture. The inevitable result is dissolution. The medieval social order is quite inadequate, and it is therefore palpably dissolving before the eyes of all beholders: religion, home, school, art, letters, intellectual life, manners, customs, and morals, the system of ownership, and the very stratification of social classes. What we are witnessing in the new freedom is not a crusade for one specific objective: the dethronement of a Catholic, or the defiance of a trans-Atlantic sovereign; it is the disavowal of an entire past. It is the dissolution of civilization—the whole of it, and the only one we have.

CHAPTER VIII

DISSOLUTION

FOR SALE—Suburban: Choice five and six room bungalows; electric lights, gas stoves, steam heat; breakfast nook in kitchen; tiled bathrooms; garages with heat if desired; close to elevated and proposed subway extensions; moderate first payment and monthly instalments.

With these sinister words every issue of every daily paper announces the break-up of the fundamental institutions of western civilization. It is not the "realtor" who is sinister, in spite of his delusive invitation to "own your own home" and let him own the mortgage. As the Official Eulogist of a great university wrote not long ago of a real estate millionaire from whose booty the university had received bountiful largesse, "He was fired by a great ideal that the people should own the land"— of course after paying him for his trouble of computing and deducting the probable increase in value of their land during the next generation or two and of erecting on it five and six room bungalows calculated not to outlast the grandchildren. But whether idealism or business, this is a common phenomenon. The realtor may be responsible for

SCIENCE: THE FALSE MESSIAH

the pseudo-tropical architectural and the match-box construction of these edifices; he may even be chargeable with the peculiar form of instalment-tenantry which suburban landlordism takes; but he is not responsible for the demand which has brought him and all his dubious devices into being. He is not the creator of the market for rented and pseudo-rented houses; for rows, and strips, and blocks of ready-made houses; for standardized houses, all exactly alike, from Forest Hills, Long Island, to Monta Villa, California, and from Bar Harbor, Maine, to Coral Gables, Florida; and above all for small houses, five and six room houses, convenient to the subway. In all this he is deferring to forces more sinister, even, than his ideals.

The full meaning of this omnipresent five room bungalow can be read only in a contrast. If we obey the exhortations of the Blue Book and slow down to twenty-five miles an hour as we drive through some old New England town, we are rewarded by a fine view of the vestiges of a civilization that has passed. These old straggling homesteads are indeed astonishing. They seem to have been set down in no sort of order ascertainable by our city-bred eyes. Apparently such a thing as a fifty-foot lot had never been heard of at the time of their construction. Relationship to rising ground, or to standing trees, was more important at that time than frontage and acreage. Moreover, few of them seem to have been built all at one time. Wings succeed wings, retreating with a succession of courtesies from three stories to two, and from two to one, and from one to low-roofed passages leading off in confusion to vague out-houses—the sheds, barns, and stables of old-fashioned husbandry. From the haphazard architecture of the successive additions, one can read the history of the family. At such a time a wing was added to make provision for the aged mother and two spinster aunts who stayed on as a permanent residue of the old family after the son and his new wife (and expected children) took possession of the original home. At such another time, the children having materialized, a kitchen was thrust back to relieve the pressure. Meanwhile prosperity came, and with it a dairy connected with the rear of the house by a passage. Such is the saga of the homestead.

It would hardly be polite to stop the car and make some excuse to inquire into the moral conditions which prevail within these premises. But that is not necessary. In general terms—and that is as far as our interest extends—they are flaunted from the gables. Like homestead, like family. We do not need to be told that this house was built by the men and women who have occupied it, often with their own hands, but always under their direction and to suit their individual needs. Probably they have never in all this history of construction and improvement considered the real estate market and whether the projected alteration would affect the salability of the house. They never expected to sell.

SCIENCE: THE FALSE MESSIAH

The successive enlargements of the establishment are still more revealing. To these people, the family meant something more extensive than a father and a mother and two children. It meant the parents of the father, if living; the mother of the mother, if widowed and without a son; any sisters of the father who remained unmarried; the children at intervals of fifteen months to a number seldom less than five or more than fifteen. This tribal group lived together under the same roof, though perhaps in different wings, for at least two reasons: the activities of the homestead were capable of absorbing them all, and there was no outside influence to separate them. Whether they lived together in perfect amity and concord is not revealed to us passers-by. And if such a question had been put to them, they would probably not have understood it. They lived together without question because there was nothing else for them to do, and the conduct of fields, barns, dairy, and house were planned about their collective man-power. They were the basic personnel of the joint undertaking. The common undertaking was possible only on a basis of the guaranteed labor of that group. Defection was high treason. In such circumstances, compatibility of temperament is a luxury not to be thought of whether it is present or not.

The five room bungalow contrasts with this establishment at every point. At the very outset it assumes the break-up of that family group about which all western civilization has until recently clustered. We may take this for granted now; but that does not mitigate its gravity. Only the most short-sighted can regard divorce as the breaking point of the family. On the contrary, it is the last despairing gasp of an already doomed institution. To be sure, there is no sacramental relation between parents and children and maiden aunts; but there is a hoary injunction to the young to "honor thy father and mother"; and that was no doubt expected to be sufficient. It always has been. Such relationships are blood-relationships, and can not be altered by swearing to love, honor, and obey. We are not told that what consanguinity has put together no man may put asunder because no man can effect consanguinity. The fundamental stability of the family system is the stability of this relationship of blood. Marriage itself derives its importance from it, since marriage dictates the channel through which the continuity of the generations is to flow. This continuity has now been broken. We assume as a matter of course that each generation is to shift for itself. The five room bungalow marks the dissolution of the ties of blood.

In his admirable survey of the basic institutions of civilization, *Primitive Society,* Professor Lowie remarks that there is no people without a clearly defined family system. The various systems differ very greatly the world over. Methods of reckoning family ties are anything but uniform. But some method of reckoning is everywhere in vogue. In every community, men, women, and children live

SCIENCE: THE FALSE MESSIAH

together in groups determined by some stable arrangement—everywhere, that is, except in modern industrial communities. There the only method of reckoning known to the European peoples is breaking up and nothing is taking its place. Of the old European family, the elders, the spinsters, and the younger brothers and sisters, have all disappeared—somewhere. It is said that between Twenty-Third and Fifty-Ninth Streets, Eighth Avenue and the Hudson River, in New York City, there lives a floating population of one million roomers. No doubt this consists of the discarded parents, aunts, mothers, and sisters of the five room bungalows.

The children are not discarded; they are prevented. Bernard Shaw once wrote that the most potent invention of the nineteenth century was the technique of contraception. But he was carried away by an exuberance of revolutionary enthusiasm. These five room manors are, certainly, birth control bungalows. They are not expected to receive the gifts of God in unlimited numbers. But before the young husband and wife have commenced to practise voluntary parenthood, they have already broken with the family system. They have left the neighborhood of all their relatives to take their "flat" or suburban villa. They have already accepted the dissolution of the family as a fact. That being the case, a limited parenthood follows as an inevitable corollary. If contraception had not existed, coincidentally with these other circumstances, it would have been necessary to invent it. Ditto divorce. The loosening of the ties between husband and wife is rather to be expected when that husband and wife are practically alone in the world. For what should they stay together? Only to satisfy our passion for not recognizing facts? If we would check divorce, we must begin by abolishing the five room bungalow. We must make it obligatory upon all married persons to remain for the duration of their lives upon the family domain of the husband; and we must provide a family enterprise which, like farming, requires to be carried on with unbroken continuity by one generation after another. Under the existing circumstances, divorce only poses this question to society: if all the other members of what was once known as the family are to be scattered to the winds, by what logic do you reason that husbands and wives can be made the sole exception? No, the five room bungalow, having abolished the ties of blood, wipes out the tie of sacrament almost without effort. It is not, as some say, the scene of progressive polygamy. That analogy does injustice to polygamy, which is a stable and highly centralized system wherever it exists. Western civilization has received no such gift. It is not abandoning one system of reckoning and taking another, like infidels converted to Mohammedanism. Its family has dissolved, and its marriage is dissolving—into chaos, contained in the five room homestead.

The bungalows are highly standardized, to be sure, and present an appearance of extreme regularity. But the regularity comes from their builders,

SCIENCE: THE FALSE MESSIAH

not from their occupants. In this they differ from the ancient homestead, which was built by the people who used it and of necessity expressed something of their peculiarities. The bungalow is regular and typical because it is ready-made. The only need it expresses is the need to move. Such houses have to be regular and typical because they have to receive a variety of tenants; and the tenants want just such interchangeable abodes because they have to move. To this end they are provided with "suites" of furniture, as standardized as their houses, planned to fit the houses—any of them—with the invariable pattern: living room suite, dining room suite, twin bed suite, and so on. The furniture, like the houses, provides the facilities only for the basic occupations of eating and sleeping. Nothing else is ever conducted in these establishments. Each house has its twenty feet of lawn in front, with a three foot strip down one side and eight down the other (if there is no garage, in which case substitute six feet of concrete drive), with twenty feet more of clothes yard at the back. On these strips grass is raised—the vestigial crop, perennial and universal, a landlord's fixture, like the plumbing. The tenant owns his lawn-mower, which he drags along with him from domicile to domicile in the safe expectation of a standardized front lawn. A rotation not of crops but of cultivators!

This rotation—the basic condition which has brought forth the lawns, the houses, and the furniture—is lovingly discussed by economists under the name of the mobility of labor. The theory is that under the benign influence of economic motives, men flow naturally to the places where they are needed, and recede in an ebbing tide from the vicinity of industrial depression. Whether this actually happens as it is described is a great question. There have been some startling migrations in the modern period not unrelated to industrial growth. In the first half of the nineteenth century crop failures in Ireland and factory building in Manchester sent swarms of men, women, and children across the channel to the industrial centers. Similarly the building of American industries has brought other swarms across the Atlantic from every country of Europe to the industrial centers of the United States. Perhaps these people have sought prosperity, and perhaps they have found it. But another way to describe their migration omits all mention of the lure that dazzled them, and points out only that great mechanical units were brought into being which demanded man-power. These mechanical enterprises accordingly reached forth and took the man-power wherever they could find it. The migrations were not tribal meanderings such as carry nomadic peoples back and forth, together with all the appurtenances and occupations of their persistently nomadic life. They were sudden transplantations. European peasants were picked up from the fields and hovels and set down in the factories and the tenements. Whatever is creative is the work of the machines.

SCIENCE: THE FALSE MESSIAH

In witness whereof, behold the five room bungalows of Gary, Indiana, and Longview, Washington. Lumbering is only lumbering, perhaps. But when lumbering is a machine industry, conducted on a large scale, and demanding a considerable manpower, the lumber industry settles upon a spot and erects a town of twenty-thousand-souls capacity over night. The five room bungalows are knocked together in rows, all drawn from the same plan, all made of the same stuff, all owned by the same firm, and all created to the same end: to house an expected population which does not yet exist. This is not merely economics; it is machinery, the reduction of human habitation to a machine product and the imposition of the requirements of machine production upon human society whatever the effect may be.

So it is with Gary, Indiana, and so it is with every factory town in every land, and so it is with modern urban life generally. The subdivision business, whereby whole blocks and boroughs are brought simultaneously under the wheels of the concrete mixer, and whole town-size suburbs raised at once in a fashion that allows the first tenant the individuality of his choice of paint, much as the first owner of an automobile can express his taste in enamel, is really one department of a vaster system ruled no less inexorably by machines. The occupations of suburbanites may be more varied than those of the residents of Gary, Indiana. They are bound together into urban areas by the inexorable requirements of transport and shipping, the

DISSOLUTION

focus of raw materials, the availability of power. These forces, mechanical forces, pile machinery together. They also pile tenements together and sow the dragon's teeth of five room bungalows in all the choice suburban subdivisions, and pile the population into them.

But the institutions of western civilization are attached to the soil (for better or worse), and refuse to pile. Consequently they are being lost in the shuffle.

PART TWO
SCIENCE PRESENTS APOLOGIES

PART TWO

SCIENCE PRESENTS APOLOGIES

As ANY one can readily ascertain simply by reading the works of scientists, there has never been a time when the scientists have not been pretty acutely aware of the effect their work was having upon established beliefs. Obviously, they have not understood its bearing upon the whole of civilization. If they had seen the process in that perspective, they would not have been troubled about science. They would have understood science itself as a symptom of a larger and deeper process. They would have seen that the driving force behind science is machine technology; and they would have understood that machine technology as a whole, and not merely the verbal promulgations of the scientists, is responsible for the dislocation of European culture. Scientists do not see this now, and they did not see it in the past. Therefore they have regarded the famous conflict "between science and religion" as a clash merely between two rival systems of ideas.

Assuming that the discrepancy is intellectual, the means of its resolution must be intellectual, too—thus the scientists have reasoned; and accordingly they have one and all, from Copernicus to Millikan, addressed themselves resolutely to resolving it. What thought hath put asunder, thought must not fail to join together. Roughly speaking, this ceremony of re-union has passed through three phases. In the first phase, science was on the defensive; in the second, the deadlock was complete; in the third, science holds the field and is dictating a victor's terms. In the first phase, therefore, the scientists were in a humble mood. We see them standing before the judgment seat of established beliefs and anxiously pleading their case with many devious arguments for their right to be allowed to live. The supreme truths—they wished it clearly understood—were revealed by a higher authority than science. With such as these they had no quarrel. The heavens declare the glory of God, the firmament showeth His handiwork; if they might be permitted to exhibit something of the richness of detail of the starry galaxy they would be indeed content, and would only too gladly join their little melody to the universal chorus of divine acclaim.

The attitude of scientists is different to-day, although it is scarcely more sophisticated. To-day we are obliged to recognize with some concern that the Almighty has gone into hiding. The voice of scepticism is loud in the land. This is a disturbing fact. No victor cares to triumph so completely that the vanquished is quite annihilated. Science now has everything its own way. It can afford to be generous; and generous it is. The prevailing

SCIENCE PRESENTS APOLOGIES

opinion is, therefore, that science on the whole endorses God. We are urged not to be hasty in dispensing with religion bag and baggage. There is much about it that is true and valuable, namely, that part which is authorized by science; and if we expect science to do the right thing by us, we must do the right thing by its protégé. Love me, love my God.

These two states of mind are not strictly chronological, of course. It is said that the science of thermo-dynamics still assumes a demonology—a divine propulsion of the particles of matter—that would have been thoroughly congenial to Copernicus or Newton. It is also said by critics who consider it a compliment that Galileo's recantation and Descartes' cautious hedging were in the best tradition of modern finger-crossing ambiguity, and for equivocation would have done credit to an American university professor. These specific charges may or may not be true. But it is true that absurd naïveties can be found in modern as well as ancient works, and that the present generation of scientific writers is not the first which has found profit in being insincere and meretricious. Still, taking the literature as a whole, it is also true that the early scientific writings approached this problem in a state of mind that was more naïve than insincere; while a wide reading of contemporary scientists does certainly produce the impression that they are—for the most part, and whether deliberately or not—decidedly ambiguous.

SCIENCE: THE FALSE MESSIAH

This, at least, is not the case with the philosophers. The central and most important episode in the relations of science and religion is called philosophy. The reason for this is that the men who played the leading parts were completely absorbed. They had neither time, energy, nor interest for anything else of consequence once they had come to realize the completeness of the deadlock between these two imposing systems of ideas. Some of them were scientists and have to their credit scientific achievements of such magnitude as to justify the invidious claim that they were the greatest intellects science has produced. Others were drawn in from the religious side; and whether like Bishop Berkeley they remained faithful to their church, or like Spinoza achieved a glorious apostasy, they compare very favorably indeed with the rank and file of their colleagues in the faith. If any evidence is required of the sincerity and candor of such men, it can be found in a little investigation which was made not long ago into the state of faith of the various professions. This inquiry disclosed among other interesting facts that of all groups the philosophers are the least pretentious about matters of belief—about which, it might be supposed, they properly should know the most. It is also interesting that those of their number who have been most celebrated for the aid and comfort they have given religion in resisting the encroachments of science and "materialism," have almost invariably not been themselves professing Christians.

SCIENCE PRESENTS APOLOGIES

Here, indeed, we have the clue to this episode in the great discrepancy, the episode of deadlock. What the philosophers have gained in sincerity and resolution, they have lost in futility and impotence. What could be more futile than to have defended religion without having been able to accept it? Others they save, themselves they can not save. This is what makes the central, philosophical, episode the most interesting and also the most difficult of the three. We are more accustomed to dealing with men who can not say what they mean, like the early scientists; and with men who do not mean what they say, like recent scientists. We have only to look over their shoulders to see what is happening to them. But the philosophers have been so resolute about pursuing their meanings to the ultimate infinitude of candor, and they have found so very little to say when they have got there and said that little with such excruciating caution, that it is always a matter of some difficulty to make out what they are talking about. But we must do it—and it is an interesting thing to do—because only in philosophy can we see how, when they are really resolutely opposed, science and religion grind each other completely into dust.

Not only have most philosophers realized this: they have also realized what it means. It means the utter helplessness of either science or religion to take the other's place; and it means the impossibility of civilization's resting on a combination of the two. The combination may be amenable to reason. It is not amenable to folk-lore. We may shut our eyes about this now, as so many scientists are advising us to do; but sooner or later we shall find it out—in retrospect.

CHAPTER IX

THE MECHANICAL DISPENSATION

IF ANY scientist or inventor, not to say president or bishop, could see in advance what changes would be wrought in human living through the use of the device he had hit upon, he would destroy it instantly and spend the balance of his life in a monastery. This not because of any special conservatism, but because there are for all men certain aspects of life that are too dear to be tampered with, and he would see his machine ruthlessly demolishing these—to his excruciating pain. But fortunately or unfortunately, we are not gifted with that clairvoyance. Therein lies the potency of the machine. If we were, it would be destroyed. Since we are not, it is allowed to remain, though we may assassinate its inventor in the depths of our folly. What the Inquisition should have done, of course, is obvious: it should have provided Galileo with a princely pension, thereby insuring his further inactivity; but it should have destroyed with the most meticulous care every vestige of his telescope. Incarcerating the inventor after his device has got abroad is precisely analogous to quarantining a man for smallpox after he has recovered from the disease—and

SCIENCE: THE FALSE MESSIAH

given it to all his friends. Galileo scarcely felt himself a heretic at all; but his telescope has been inciting other men to heresy for three hundred years. Heresies are not matters of faith; they are common observations. The church was right in its judgment of what Galileo did, and wrong about the importance of what he said, because what he did could be and was repeated. The only effective heresies are the ones that arise in this way—as the result of general practises which gradually lead men to new ways of thinking. Such states of mind can be heretical, of course, only if the new practises disturb the ancient order of things. The state of mind of any people is after all part of its total adjustment to life. Ideas are habits no less than institutions. In any stable civilization, the ideas and the institutions, and the means of livelihood and the physical environment are in pretty complete adjustment to each other. That is what civilization is: a scheme of adjustment. Anything which upsets that adjustment will necessarily cause a long train of readjustments which will begin with the aspects of life that are nearest the impinging change and end with those which are most remote and general. In the present state of history and anthropology no absolute dogma is possible. Yet there are many indications that such punctures in the established order of civilization occur almost by necessity in the barnyard rather than the temple or the capitol. So long as the means of livelihood of a people remain essentially unchanged, its civilization is

THE MECHANICAL DISPENSATION

likely to weather even fairly cataclysmic changes in the personnel, or constitutional superstructure, of the body politic. But a change in the weather affects every stomach, and consequently every heart and head, if the change is great enough and permanent enough. The only heresies that are effective are the ones that begin at home.

This is why the heresies of the machine are so superlatively effective. Machinery begins by altering the day's routine, and ends by altering the cosmos. Its effects are universal because its influence is universal. Nothing, indeed, but the forces of nature is satisfactory for comparison with it. The effect of a technology which multiplies by one hundred the population which a given area, say England, is capable of sustaining, is the same as though the area of England be conceived as extended a hundred times. But such a conceit like all fancies based upon magnitude alone throws the imagination off the scent. To conceive modern industrial England as medieval agricultural England multiplied by any given figure is to lose the point of contrast. It would be far more realistic to note that the area of England, or any other country, has actually shrunk by as much as the difference in the rapidity of transport will account for. Here is a comparison which is not imaginary. It is not "as if" the world were changed in magnitude. For human purposes, the world is actually changed.

The significance of machinery for human civilization lies in the fact that it alters what is most

SCIENCE: THE FALSE MESSIAH

fundamental to man: topography and climate. It makes regions accessible which were never accessible before. That is an alteration in topography. It makes wheat grow where it would never grow before. That is an alteration in climate. Even this figure is not quite adequate. Machinery has altered the significance of earth to man. Whereas all human civilization in the past has been bound to the soil; now it is indentured to the machine. The difference does not appear important at the breakfast table. What is consumed there is wheat and corn, eggs and milk, berries and fruit. Biologically, man is the same animal he has always been. He is totally dependent for life upon the exploitation of plants and animals. Furthermore, man has never been a grazing herbivore or beast of prey. As a hunter, a herder, an agriculturalist, he has always used tools. That means, of course, that there has always been some degree of division of labor, some arts and crafts in every human economy. But this is not the point. The transformation effected by machine industry is a re-creation, not of digestion nor of manual skill, but of community life. The requirements of the stomach are one thing. Conditions of life are another. The machine has altered the latter. Whereas in the past, not only man as a species, but every community as an organized group of men struggling to wrest the means of life from the jungle or the soil, have been bound to the soil or the jungle upon which they were located; now tremendous numbers, forming not only cities

THE MECHANICAL DISPENSATION

but even nations, more populous than ancient empires, exist more or less independently of the food-bearing properties of the terrain on which they live; and as time goes on the dependence is less rather than more. There is no need to enlarge upon the situation. The facts are obvious.

What is not obvious, though it is incalculably important, is the effect of the change. That effect is mental. The imagery in which men envisage their thought is taken from their immediate physical environment. Where the community is agricultural with only a scattering of artisans to punctuate the prevailing rusticity, the physical environment that infolds men's thoughts is that of the countryside, pasture land, or jungle. For the industrial community, whatever it may receive instead, that setting is removed. The physical environment of human life as it appeals to the imagination of common men is altered by urban life and machine employment. Machinery alters the environing climate and topography by removing wind and weather from human preoccupation and substituting something else. Entirely apart from what that something else may be, this alone is a profound change.

For one thing, the climate of wind and weather is utterly precarious. Without denying that industry is subject to fluctuation and that hard times do recur under industrial economy, we can nevertheless see that in exempting men from the utter day to day precariousness of storm and drought, ma-

SCIENCE: THE FALSE MESSIAH

chinery has effected a complete reorganization of human imagination. The weather has been withdrawn from circulation as the pattern of human thought, and with the weather—the Great Spirit. Recall the Litany: "From lightning and tempest, Good Lord, deliver us." Take note of the common law. In the common law the vagaries of climate are definitely personal. In the receipts issued by express companies, and in insurance policies, earthquakes and tornadoes are described as "acts of God" as a matter of course. Because these cases are closer to us than the rain gods of primitive man they make a very ancient fact most vivid. The forces of nature have always figured in the dreams of men as personal agencies. In short, animism has colored all human thinking, from the simplest savages down to the authors of the *Book of Common Prayer.*

The usual conception of animism is, of course, that it is a projection of man's picture of himself into the impersonal realm of nature. But that definition is too sophisticated. Where, then, did the picture of human nature come from, that it is thus projected? Is primitive man to be credited with psychological insight beyond his mastery of natural history? If the savage makes any mistakes about nature, he is just as likely to make some about his own nature as about the outer world; indeed, he is certain to. Whatever he thinks of one he will think of the other, not because he projects a principle he understands farther than is exactly right and

THE MECHANICAL DISPENSATION

proper, but because he is a natural monist (consistency being, apparently, a feature of the cerebral cortex) and feels impelled to apply much the same imagery to all the subjects he considers. Animism is not a projection of a "presumably true" conception of the human spirit into the external world. It is a conception of both the world and man in certain terms. What those terms have been is obvious. How remote they are from the scientific view of the human mind modern psychology is now showing, just as vividly as modern astronomy ever exhibited the short-comings of animistic star-gazers. The trouble with the common definition of animism is that it was laid down by those astronomers before the scientific psychologists had come along.

What probably happens in the mind of the man who communes with nature, as all human civilization has done heretofore, is that he assigns to the winds the emotions natural to man, and to man the powers natural to the winds. The great spirits have feelings like man; man is a causal agent like the spirits that rule the wind and the wave. The sense of cataclysmic force is read into the entire cosmos evenly. Man's power is a matter of overweening pride; nature's is one of blind terror. But in either case the emotional tone of the conception comes from the abject dependence of man upon the weather. When the sun shines, the grain shoots up. When the wind blows, the tender sprouts are all laid low. The sense of power, of a power that is beneficent or malevolent as all human operations are, is

SCIENCE: THE FALSE MESSIAH

the leading motive in the whole symphony of human thought. The one thought which is perpetually hovering in the human imagination is that of the relation of man to the powers that rule his destiny. Animism is not the product of intellectual naïveté. In one form or another it is the philosophical embodiment in human thought of man's relation to nature whenever and wherever he lives upon the soil.

But the precariousness of life is not the only reflection that arises from dependence upon the soil. If man is at the mercy of the Forces of Nature, the grains and the flocks are at the mercy of man. They, moreover, are organisms whose life cycles very closely resemble that of the human body. The interchangeability of man and the domestic animals or even the domestic plants is obvious. This reflection also gives meaning to the animism of domestic man. The inevitable disposition of the farmer or the shepherd is to treat man, plants, and animals very much alike. What happens to one, happens to the others. What is good for one, is good for all. Reproductive potency, upon which husbandry depends, as well as human society, is much the same phenomenon in every species. Human fertility is symbolized by the overflowing cornucopia; the multiplication of the cereals is induced by rites of human sexuality, and is represented even to-day by the ripe figure of the vigorously procreative woman. Mutual identification is complete.

THE MECHANICAL DISPENSATION

This identification is a product of common labor in the fields. Only by the fusion of man and plants and animals in the toils of the husbandman does the imaginative fusion take place. The three are mutually assimilable only as the object of a common solicitude. To careless and indifferent people—children, for example—they bear no resemblance to one another. Nothing about a flowering plant suggests a parallel to the human generative cycle, unless the cycle of the plant be followed through with acute solicitude for maximum fertility and a bumper crop. In this sense, also, the insistent animism that runs through the whole history of human thought is an expression of the common occupations and common interests of men who have lived upon the soil.

The onset of mechanical civilization is the great divide. It is not a meander in the rivers of civilized life in which earlier confluents are led over another course. It is another watershed. No one can begin to grasp its significance until he has realized this. The beginning of modern thought is not some discovery that has turned the current of the prevailing state of mind. Its real source is rather the first trickle of the imagination, the attitudes and ideas, of common men who live at the head of the mechanical watershed, on the modern side of the great divide. The state of mind of modern civilization is not a stage in the history of science. It is the realization in articulate form of what it means to live in the world science has helped invention to create.

SCIENCE: THE FALSE MESSIAH

The course the waters will take is determined by the character of the watershed. The philosophy of the next generation is perceptible now only in the condition of life of the common people. From them flow the rivulets of imagery.

Nevertheless science has appropriated the credit and the blame for this event. We have found ourselves thinking less of divine guidance in recent years and more of mechanical causation. To some of us the change has been natural and agreeable, or at least tolerable. We have accepted it without thinking much about it. But not to all: those who have had a stake in the old order have objected vigorously; and finding us offering up burnt offerings to the golden calf of science, they have seized upon it as the cause of our defection and at least attempted to trample it in the dust, overlooking the subtler causes of our change of heart something as Moses overlooked the part of Aaron in the celebrated episode in which, as Exodus XXXII: 35 has it, "The Lord plagued the people, because they made the calf, which Aaron made." Thus science has been plagued for doing what has been accomplished by other and subtler agencies. All scientists have resented this treatment and many of them have actually deplored the social changes of which science has seemed to be the cause. Thus we have been presented with the extraordinary spectacle of scientific apologetics: the attempt to relieve science of the odium of a crime in which it has been at the most only an accessory after the fact.

CHAPTER X

SCIENCE MEEK AND MILD

"I BELIEVE in atoms, molecules, and electrons, matter of heaven and earth, and electrical energy its only form. I believe in modern science, conceived by Copernicus and borne out by Newton, which suffered under the Inquisition, was persecuted and anathematized, but rose to be the right hand of civilization as a consequence of the fact that it rules the quick and the dead. I believe in the National Research Council, the communion of scientists, the publication of discoveries, the control of nature, and progress everlasting. Amen."

This is blasphemy. To put our scientific opinions in this form implies that we have put our belief in atoms, molecules, and electrons in the place of our faith in God the Father Almighty, Maker of Heaven and Earth; and very few of us are prepared to do exactly this. We see no reason why we should. Science to us is nothing more than a collection of facts. Electrons and molecules mean very little to us except as curiosities, like Egyptian mummies and dinosaur eggs. The things of science are just such things as can be seen in the museums, very interesting and in-

SCIENCE: THE FALSE MESSIAH

formative and of course all definitely proved, but neither shocking nor inspirational. They are just facts, and facts do not touch us very intimately.

No doubt this state of mind is naïve. But it was the state of mind of most scientists at least until quite recently. The early discoverers of the planetary orbits and the motions of the earth regarded such matters merely as curiosities. They expected nothing in particular to come of them. Their imaginations were by no means wholly dominated by the imagery of colloids and Brownian movements as the imagination of primitive man was dominated by the Great Manitou, or medieval man by the Blessed Virgin. This is what scientific writers mean when they emphasize the importance of idle curiosity in scientific work. Most scientific discoveries, we are told, have been provoked by sheer intellectual curiosity. That is, their authors have not been spurred by the hope of any sort of gain to themselves, either spiritual or physical. They were not intent upon making their fortunes, and certainly had no thought of discovering a "northern passage" to the salvation of their souls. They expected whatever they might find, if they should be so lucky as to find anything, to be just an interesting curiosity. If we think of the truths of science merely as so many facts, overlooking for the moment their importance for mechanical technology—and that is the way scientists have taken them—they are just so much intellectual bric-a-brac. What we think about atoms is no great concern to us as men, and what we believe or disbelieve about colloids or catalytic agents is inconsequential because such things do not matter very much in our lives. We are under no overpowering necessity of making peace with molecules.

It is different with spiritual truth. We are indeed called upon to "get right with God" through the peace that passeth all understanding. The spiritual truths to which we turn for help were not deduced out of idle curiosity, and later found to be beautiful and helpful. We needed them, presumably, before we got them. The Great Spirit was not imagined first by some primitive genius, and afterward set up in contrast to the precariousness of life and worshiped as a means of providing against it. The condition existed before the philosophy. Some theory of life was required because men lived and had dealings with the precarious world governed by the Manitou. Wind and wave, cloudy sky and brilliant sun, and the magically benignant warmth of spring, the stirring of the sap and the recurrence of the procreative urge—all bespoke a universe with which human beings must somehow manage to live at peace. Theologies are not sheer wayward ingenuities. They express what life requires. And what the life of man has always required (heretofore) has been a conception of the universe in such personal terms as should make it a working partner in the personal activities of human beings.

Science has never intended to alter this relation-

SCIENCE: THE FALSE MESSIAH

ship. It has no concern with it. Neither has it intended to displace "spiritual truth." It is not a competitor with spiritual truth for man's affections. If the two have come into accidental conflict because a certain belief of science has occupied the same field as a certain older belief of theology, this has never meant that science intended men to treat the scientific belief as a beautiful and helpful truth. Science has always hoped and prayed that theology might be just as helpful without that particular belief, and that science might be allowed to state the truth in that matter without being thought presumptuous.

When science first appeared over the horizon, the older folk-lore may have felt disgruntled. Captious critics within the Roman Church eventually noticed that Copernicus had usurped their prerogative of issuing all the guide-books to the heavens and had reversed some of their favorite itineraries. This was cause enough for reasonable annoyance. But the annoyance was short-lived. It soon appeared that even though the orbits of the planets were as Copernicus had indicated, no real trespass had been committed within the pearly gates. Indeed, the Copernicuses and Galileos have always been quite ready to disavow any such intention. Therefore, after a few moments to think the matter over, the churches have always been able to accommodate themselves to each new scientific promulgation. The prophets of both departments of the folk-lore made it up after each new unpleasantness with joint announcements that there is no real conflict between science and religion.

They are right. Any body of scientific lore in any one of our sciences can be assimilated to the rest of our folk-lore with comparatively little strain, and for that matter has been. The Copernican system of astronomy seems to us to ruin the whole Christian conception of Heaven and Hell, and of the central position of man in the universe and the esteem of the Lord of the universe. But in the sixteenth century adjustments were made without difficulty. After all, Copernicus laid down no theory as to the location of the hereafter; that could be left to the Christian folk-lore as theretofore. Similarly, Newtonian physics had the effect of ruling completely out of the universe any manifestation of divine power in the form of an interference with the even flow of physical action and reaction. This might seem to be a death-blow to the Christian law of divine government and guidance; but it was nothing of the kind. A supreme and omnipotent Being must of course be able to conduct a universe without the backing and filling to which men are accustomed. He would—it now seemed obvious—by preference design it so perfectly that forevermore it would proceed under its own steam. His management and guidance would thus be implicit, put into the whole system at the start and effective throughout every operation. Thus the very laws of motion were only so much more evidence, quite desirable evidence, of the perfection of God's design. So also

SCIENCE: THE FALSE MESSIAH

evolution: God created the universe in six days which we are now to understand as six tremendous periods of time. This is more plausible than six literal days, and is altogether a desirable amendment of the Mosaic law. Man was created last, as evolution and Genesis agree, and a little lower than the angels. That this means also a little higher than the monkeys is a point of no theological concern. God moves in His mysterious ways His wonders to perform.

But such an account makes it appear that all the assimilation has been in one direction. That is not the case by any means. Every scientific theory has been put together out of the materials at hand, and these have been in large part supplied by the lore of theology. Copernicus worked out a system of planetary relations quite different from the one in general vogue; but he had no notion of representing his scheme of the relation of sun and earth and stars as anything else than the divine order in which these bodies were placed by the Almighty. That the position of the planets in the heavens was decreed by the Almighty was axiomatic in his mind. So also with Newton: his theory of gravitation called upon distant bodies to draw near to each other. How could distant objects be supposed to answer in this fashion? Clearly they were compelled. Newton's idea of the compulsion which ruled their movements was precisely that of a good deist. He supposed that it was part of the divine plan that they must. The exertion of force at a distance was conceivable

SCIENCE MEEK AND MILD

to him only by the help of his religious lore. And evolution: how could it ever have arrived at man without the guidance of Omnipotence? The weakness of evolution in Darwin's time—and this is still its weakness—was the absence of any sound clue to the forces which bring about the modifications of the species. The hypothesis accounts for nothing unless we help it out with materials from another source. Darwin was perfectly ready to do this. So was Newton. So was Copernicus. So are all the scientists. They must be. Otherwise their theories are inconceivable. If science would only come to rest at a given point, the assimilation would soon be quite complete, with the scientists and the theologians in full agreement, and all lying down together in the best millennial convention.

Indeed, no one is more anxious than the scientist for just this to occur. We ordinarily think of scientists as men who are fired only by a zeal for science and its advancement. But scientists were men before they were scientists, and sometimes are after. They would be only human if they were most zealous for the advancement of science chiefly at the particular spot at which their own researches fall; while they would be distinctly more than human if they did not deplore sincerely and heartily the unsettling effect of certain types of "theories" upon civilization. Mathematicians are usually God-fearing men. Civilization rallied from the effects of modern mathematics in the time of Descartes, and very little effort at mutual assimilation has been

SCIENCE: THE FALSE MESSIAH

necessary since. Perhaps they have not rested entirely content with the religious folk-lore of that time; but they have not been the restless ones, champing at the bit, and demanding to get out of the rut. And by the same token they may be pardoned if they have been a little impatient from time to time with biologists, or psychologists, who—not entirely content with Descartes' doctrine of the relation of the central nervous system to the soul—have been trying to improve upon it, and so have been running grave risks of upsetting the apple cart all over again. Many a physicist and astronomer has been sorely provoked in recent years by the disgraceful light-mindedness of psychologists. They fought and bled in the sixteenth and seventeenth centuries; recovered from it; and decided that there is no conflict between science and religion after all; and they do not thank these modern theorists—their ideas are so vague that they can hardly even be called scientists!—for reopening issues which had much better for every one's peace remain closed.

This is all well and good for Newtonian physics and Copernican astronomy. The established folk-lore of science consorts with the folk-lore of religion as peaceably as any one could wish. The difficulty is that the folk-lore of science does not stay established. Consequently, no safeguard against heresy can be effective which is not general and progressive. What civilization needs is protection against every conceivable scientific theory, a reconciliation between all possible religion and all possible science. Such a general reconciliation must be couched in very general terms—philosophical terms. It is likely also to result in a reconciliation so general, so philosophical, that for those who succeed in making it at all, Reconciliation becomes an end in itself in which both science and religion have disappeared.

CHAPTER XI

MAKING RELIGION SCIENTIFIC

ONCE a man has begun to think scientifically, his religion invariably commences to trouble him. He may be able to dispose of the facts of science. He may be able to stow away the separate truths of science and theology in non-communicating cells of his imagination. But this does not reconcile their methods, which still remain to plague him by their continual reminder that however he may arrange his various beliefs he is nevertheless receiving them in two sets of mental processes which seem to differ widely. Scientists usually call this the difference between demonstration and belief; and theologians contrast the evidence of the senses with the evidence of the human heart. Science is of the senses, demonstrable; religion is of the heart, revealed. But neither of these descriptions is quite satisfactory. Science is as much a matter of belief as any creed, and religion is no more revealed than anything else except in the apprehension of its devotees. Each is demonstrated according to its nature.

This is where the discrepancy occurs. The nature of the demonstrations differs widely. Science is demonstrated by instruments; religion by rhetoric, or, if you like, by poetry. In each case the method and its resulting truths are peculiarly adapted to each other. The facts of science are peculiarly mechanical, and the verities of theology are peculiarly rhetorical, or poetical. The arguments of science are rigorous. They compel the assent of our intelligence, which seems to be peculiarly susceptible to mechanical appeals. The arguments of theology are sentimental. They proceed by blandishment. Their appeal, as theologians declare, is to our hearts, or to our emotional temperament, which likewise is most susceptible to the charm of ritual and liturgy.

Which of these two aspects of our being is more reliable it is not within our province to decide. Why should we assume that either is particularly trustworthy? We do; but that is another matter. For the most part, and taking the whole history of civilization into account, we have reposed our trust chiefly in feeling and temperament. The appeal by which civilization has prevailed has been the appeal to sentiment and not to reason. Folk-lore is the repository of our sentiments. That is its nature. But once the intelligence has been aroused, once we have begun to cultivate our mechanical facility, we find it increasingly hard to recline at ease upon our prejudices. We want to make everything mechanical. We want to analyze our feelings and demonstrate our sentiments. This is bed rock of the famous controversy. If our history had been different, perhaps it might have taken the opposite

SCIENCE: THE FALSE MESSIAH

form of a theological revolt against the sciences; or perhaps that may occur as an aftermath of the mechanical Utopia—after an era of intelligence, the re-discovery of the heart! That would be poetic justice. In the present instance, however, we are discovering the intellect, and having carried it to the utmost limits of mechanical ingenuity, we are very naturally trying to reduce all the aspects of our life to its sharp visual clarity and its mechanical precision. We would now justify belief by demonstration!

Clarity and precision were not wholly unknown before machinery. There is a certain clarity in folklore. Whatever is thoroughly familiar is felt also to be quite clear and sure and even quite precise. What do these conceptions mean, in ordinary speech, except that we are satisfied with our knowledge of the subject under discussion? There is no general standard of precision, fixed by the stars, nor even a customary norm, fixed by folkways as the standards of weights and measures are. The only norm is in our heads. Every man has his own notion of clarity according to the translucence or the cloudiness of his mental processes. To agnostics, the arithmetic of the Holy Trinity is so much nonsense: it appears impossible for three ever to be one. To Utopian revolutionists the arithmetic of husband and wife, the unity in which two separate individuals figure as a single entity by religion, in law, and under the usages of social custom—"the two who are two and yet are one,"—is incredible hocus-pocus. To hard-headed farmers, the mystic unity of capitalists in a corporation, a legal personality which can be sued, convicted, and pronounced guilty of the most flagrant offenses against common decency without in the least diminishing the good standing of any of the individual men who compose the corporation, is a palpable deceit and a conspiracy against "the farmers." On the other hand, farmers are thoroughly familiar with marriage, and if any one should suggest to them that the union of man and wife by ecclesiastical benediction is a haze of popular mythology, they would regard that man as slightly daft. Meanwhile the coupon-clipping cynic who thinks nothing of laying his sins upon the shoulders of a legal fiction may be quite "disillusioned" about marriage and still take the abbreviations in his stock reports as the names not of fictions but of actual and undoubted realities, as certain and precise in all their outlines as—well, perhaps, as the Holy Trinity. There is no contradiction here, though there may be some confusion. All these opinions are held on excellent grounds. What people mean by precision in all ordinary dealings, is familiarity, and nothing else whatever. In this fashion, every wife understands her husband perfectly, and may even penetrate the ways of God.

All this is commonplace. But what strikes the eye in the engine-room is a different sort of precision—a perfect precision, if you like. Clarity of this sort is not a product of long association. It is

SCIENCE: THE FALSE MESSIAH

perceptible at first glance. Hence its fascination. From the moment of first inspection, it is perfectly clear that the piston will not rattle, the crank-shaft will not get out of step with piston and connecting-rod, though the engine turn over at the speed of five thousand revolutions per minute. Similarly, one careful examination of an isosceles triangle is enough to convince even the most sceptical that the angles made by a perpendicular from the apex to the base will always and forever be equal angles, regardless of every circumstance of size and position. Here is a certainty to give one pause!

Those early mathematicians and astronomers who devised the first isosceles triangle and the first ratios of the movements of the planets are not to be blamed for having been fascinated by their own discoveries. What they had devised was so rare, in a civilization unblessed with steam and electricity, as to be a prodigious curiosity. The volume of scientific lore was in their day quite infinitesimal in comparison with the hurly-burly of common life. But it sparkled as a diamond. Rare as it was, it could yet so impose itself upon the imagination as to convince those Chaldean astronomers or Pythagorean mathematicians that the essence of the universe must be such a rigorous precision. How much more, the witnesses of the birth of the solar system and the differential calculus! Only in the very best good faith did the thinkers of modern times set out to make the confusion of the folk-lore "conform to the dictates of intelligence." How clear might not our hopes and fears become, reproduced in blue-print in perfect scale and with every detail plainly noted!

In geometry, the most gratifying results have been achieved by the application of what is now called machine standardization. That is, all the elements of the argument are reduced to interchangeable parts. Each axiom and theorem is stated once for all in a polished formula of words that remains for ever of the same gage and can be introduced into the argument whenever its gears mesh. The demonstration of the more advanced propositions is precisely like the building of a complicated machine. Every part that goes into a gas engine, for example, is by itself quite simple. The subtlety of the assembled engine, like that of the Pythagorean proposition, or of Napier's analogies, is wholly one of assembly. If the gears mesh, the engine will go, or the problem will solve.

To obtain an eternal and immutable system of morality, or to demonstrate the existence and nature of God—and so to give the folk-lore all the driving-power of machinery—it therefore seemed necessary to seventeenth and eighteenth century philosophers only to apply the principle of machine standardization to ethics and theology. The highest good and the most perfect Being, high and perfect as they are, could hardly be conceived to be more complicated than the contrivances of science. Perplexing and anomalous, yes; but not extensive or highly fabricated. The application seemed quite feasible;

SCIENCE: THE FALSE MESSIAH

and it was accordingly made, and resulted in a considerable literature of works that are mathematical in form and contrivance, but spiritual in content and intention, a curious literature from which we still pluck an occasional apple of ripe worldly wisdom, but by which in the end nobody seems to have been convinced.

For this spiritual geometry axioms are necessary, and other parts. Note that the whole secret of the clarity and precision of machines and mechanical reasoning lies in the simplicity of the unassembled parts. The first task of metaphysics, therefore, is to disassemble the various constituents of folk-lore, using universal doubt as a mechanic would use a universal wrench. The whole structure being thus unhitched, reconditioning along scientific lines can then go forward, beginning with some basic part that has been stripped to irreducible simplicity. In the most famous demonstration of this sort—Descartes'—the axiom chosen was the fact of doubt. This seemed elemental and indubitable, a satisfactory shaft to which to attach the gears and driving wheels that set the engine going with a restored belief in God, freedom, and immortality as automotive conceptions, fully tested and guaranteed to run.

But ancient and worn-out parts have a way of refusing to mesh. One by one Descartes laid his *disjecta membra* in a rusty row under the hood of his machine, and when they were all accounted for he clipped down the hood and expected them to carry the load. And so they did—for him and for all the rest of his contemporaries who had never seen an engine in motion and could hardly be expected to know that Descartes' machine was a strictly non-functioning wreck. They knew what was under the hood: I doubt—therefore I am—therefore I am a thinking being—therefore I am able to conceive the idea of a God—therefore God exists precisely as I think Him, an omnipotent and perfect Being. The unmechanical character of such an array is so patent that to-day (when mechanics are better understood) it is noticed by all the critics and has even been made part of the stock in trade of elementary instruction in the subtleties of metaphysics. As the professors say, in Philosophy I, Descartes has implied his conclusion in his premise. That innocuous "I doubt" does not, after all, mean that "Nemo functions according to the unknown capacities of a being as yet specified only as a pure variable." On the contrary, it means, just as it says, that Monsieur René Descartes is registering the scepticism appropriate to a gentleman and a mathematician, even in the seventeenth century, and thereafter it is with little surprise that we discover that the same Monsieur Descartes believes in God, he having been educated in a Jesuit school.

What professors have not noted, however, is the fundamental discrepancy between items drawn from mathematics and items drawn from the common suppositions and beliefs of folk-lore. Descartes presumed that the key-note of mechanics is clarity

SCIENCE: THE FALSE MESSIAH

and precision, and so it is. He also believed that his notion of doubt, or even deity, was perfectly clear and precise. But he failed to distinguish two quite various standards of clarity: one the machined smoothness of parts which have been ground to mesh, the other the familiarity of seasoned acquaintance and old affection. So it is with the whole of spiritual mechanics. The principles of mechanical reasoning are indeed as drivingly cogent as bevel gears; but they are so only when the materials thus driven are indeterminate numbers and pure variables and readings from the dials of machines. Only such things mesh with each other in this fashion. Folk-lore—in which two can melt imperceptibly into one after the repetition of appropriate ceremonies—is not the stuff of algebra. A misfortune for algebra no less than for folk-lore!

But when its center of gravity is shifting a civilization must be prepared for such misfortunes. The mathematical demonstrations of the existence of God, and the search for the fundamental axioms of theology have gone somewhat out of fashion, though they are still to be encountered in the literature of the obscurer and more esoteric religious cults. But the necessity for them has remained. Consequently, though styles have changed, "proofs" are commoner than ever. Indeed, so common are they that few realize any more how peculiar they are to our predicament. It seems perfectly natural to us that spiritual things should be subjects of demonstration. We may not accept fully the logic of these exhibitions that the parting of the Red Sea was a volcanic phenomenon and that the devils were cast out by the technique of psycho-analysis; but we listen to them without smiling, thinking them quite the sort of thing we must expect if folk-lore is to conform to intelligence. It never occurs to us to inquire why it should. Yet in what other civilization have the gods been represented as possessed of a scientific education and the powers of qualified practitioners of the mechanical arts? One of the favorite theorems of modern apologetics shows that God is no more mysterious than electricity. Indeed! Only electricity, then, is convincingly mysterious to the modern mind!

This, of course, is the fact, and the source of all our trouble. Such are the conditions of our life that it has become necessary to gage the horse-power of the Omnipotent. If this demonstration is ineffective, then it will be necessary to go further, and prove the unreality both of horses and of power. Retreat to the age of faith seems to be out of the question—just as retreat to the age of handicraft. Whatever the absurdities, all the contents of our folk-lore must be subjected to the dictates of intelligence, which is to say the dictates of machinery. Whether we have any folk-lore left in the end depends not upon our need for it but upon the requirements of machinery.

CHAPTER XII

SCIENCE UNTO CÆSAR

PERHAPS the favorite device for reconciling superstition and enlightenment is the system of benevolent neutrality, otherwise known as rendering unto Cæsar. If we are practical men, we honor the things of the spirit but let business be business. If we are scientists, we believe what can not be proved and prove what can not be believed. Thus we make a dualism and call it peace.

Dualism, of course, is no new thing. If there was ever a beginning to the discrimination of sacred and profane, the archeologists have not found it. However primitive a people may be, none is so poor as to lack a heavenly home, or so base as not to own a soul, to bury its dead facing east with a coin upon the tongue to pay the ghostly boatman's fare, or perhaps in full panoply of war or hunt and with all the accouterments of rank so that the departed will not be without means of subsistence or unable to maintain his dignity and station in the Elysian fields.

For some reason, these distinctions of sacred and profane, soul and body, heaven and earth, are always turned to the disadvantage of mundane appurtenances. What this means is doubtful, like the bearing of ancient laws forbidding usury or the coveting of a neighbor's wife. Such laws can be interpreted as showing the purity of those peoples who so explicitly discountenanced what we also regard as sin. Or they can be interpreted as an indication that prohibitions of these things were urgently required. Possibly all civilizations, including our own, have placed spiritual things upon a higher plane and have gone explicitly on record as so doing. Possibly, however, the universal disposition of mankind is to take thought first of the day, what we shall eat and what we shall drink, so that the prophets of all the ages have been obliged to restore the balance by crying, "Woe, woe, idle and corrupt generation!" and so have filled the literature of the world with misleadingly sententious documents. However this may be, our literature and traditions are replete with discriminations of this kind, ostensibly bearing right hard on what is merely present and human and agreeable.

To give this perennial dualism its peculiarly modern form, we have only to apply it to knowledge. The ancients distinguished objects sacred and profane: for instance, the sacred bull from the profane pig. We have been able to "understand" that such a distinction "really" applies not between two sets of objects in the same perspective, but between two perspectives. In one the entire universe is sacred; in the other the same universe is entirely profane. The law of conservation of energy (which

SCIENCE: THE FALSE MESSIAH

holds that all physical processes are continuous and uninterrupted) is a profane picture of the universe. The law of immanence (which holds that God is everywhere expressing His divine will in all things) is a sacred picture of the same universe. Our modern theory is that neither of these perspectives in any way detracts from the other. We render unto science that which is science's, and unto God that which is God's, and congratulate ourselves upon the subtlety of our insight—not the least triumph of this, the highest of all civilizations!

Although this subtlety was originated by some of the keenest minds of modern times, the process by which it was attained is not difficult to understand except in its historical ramifications. In principle, it consists only of applying the standards of animism to science and observing the incapacity of science to measure up to the test of folk-lore. The subject-matter of science is instrumental and mechanical; whereas the proper study of mankind is man. The human imagination—uncontaminated by machines—is concerned primarily with agency. Where any human problem is involved, our sophisticated maxim is *"Cherchez la femme!"* This includes *"Cherchez l'homme,"* and, as another case of the universal search for agency, *"Cherchez le Dieu."* Like man, like God. Whatever happens, some one human or quasi-human is behind it. The problem is to find who it is, what he wants, what he is trying to do, and how to stop him. Folk-lore is not preoccupied with psychological subtleties.

Issues of free will and determinism cut no ice for Eskimos and no bamboo for Borneans. When the agent is discovered, he is treated as a unique and unfathomable spring of motives and actions.

In folk-lore, accordingly, cause means agency. The pursuit of conditioning causes for any phenomenon is a search for agency: *Cherchez l'homme.* This is true not among primitives alone, but among all men. What an agent may do varies, of course, from one civilization to another. Among Zulus, it doubtless seems quite reasonable that one man should cause the death of another by burning his nail-parings with appropriate incantations, just as to some modern Americans it seems possible for one man to cure another of disease by sitting in his study at home and reading out of a book. If the Zulu dies, his friends and relations pay less attention to the details of his basal metabolism than they do the question who caused his death; while among us, no one is ever allowed to have recovered without some one having cured him. Perhaps it would be an exaggeration to say that the medical profession depend for their employment upon the popular belief in agency; but surely the manner of their employment is formed by that belief. All of us have often heard it said that we can see what science has done for us if we will only reflect that a thousand years ago when a man took cold he attributed his misfortune to the influence of the devil, while a man who takes cold to-day knows that he got it because he neglected to wear rubbers. This is an

SCIENCE: THE FALSE MESSIAH

interesting reflection. It is splendid, of course, that we are able to know the cause of things, whereas our ancestors could at best only vaguely suppose the demons who influenced their lives. But somehow the theory of agency has managed to remain intact. In one age the cause of disease is a malicious spirit or an angry God; in another it is a stubborn man who declined the precautionary ritual. What an emancipation!

Now every one who has performed or attended autopsies knows that the examination of a dead body usually reveals at least half a dozen adequate causes of death. If the condition for which the patient was being treated is revealed as among those present, it is usually assigned as the official cause simply for legal and professional convenience. But science is no such respecter of persons. Science reveals a varied scene of dissolution and mortification which can be traced back only to an equally varied and complicated array of preceding conditions, each with its long line of earlier states, predisposing causes, and aggravating accidents, until it becomes clear in very truth that the man has spent his life a-dying. To this potpourri of causes and effects, the notion of a responsible agency is utterly irrelevant for two reasons. The designation of an agent implies that it was amply able to produce the whole effect by its own prepotent power. No such agent exists in science. It also implies that the search for causes comes to rest with the discovery of an agent, since the agent is what we are looking for. But in science the search for causes never comes to rest and there is no preconceived dogma defining the satisfactory conclusion of research. In science, one research leads always to another. The effect of one discovery is only to exhibit the imperative necessity of many more if what has just been discovered is to be understood.

This is not exactly the course of the reflections of David Hume; but what he wrote came to much the same thing, so far as science is concerned. Hume took the "category" of cause—so called in metaphysics because it is a basic element of scientific thinking—and with superlative nicety dissected it. When the dissection was complete, and all the constituent meanings that enter into the causal relationship were lying on the table, Hume found that they were the succession of events in time, the close association in temporal sequence of one event with another, the vivid resemblance of causes and effects, and the familiar grouping of two or more events in a succession frequently observed so that the presence of one inevitably suggested to the imagination the presence of the others. In short, cause and effect mean nothing more than the earlier and later stages of the same condition. For the infinite mind one might suppose that all the world forms one vast condition, and that causal relations would be inconceivable except as the whole flow of world events, each stage of which follows the other simply by virtue of following—if stages could exist for such a mind. But to the limited perceptions of

SCIENCE: THE FALSE MESSIAH

human observers, only partial views are possible. When, therefore, a man inquires what is the cause of this or that, he means "What has preceded this condition as part of the same fragment of the world?" His choice of fragments is wholly a matter of circumstances and his human limitations, and a slight shift of circumstances will send him off on another line sedulously tracing the succession of things on some other field of valor.

The whole argument can be summarized in this fashion. Causation is the succession of events. As all things occur together in this world, the selection of specific causes of specific events is arbitrary. The events may occur in the sequence noted; but as each is only a part of a general condition in spite of its arbitrary selection in this research, their designation as "the" cause of "this" effect is arbitrary. Any cause can be related to other effects, and any effect can be related to other causes, since each is followed or preceded by many more circumstances than the ones noted.

All of which comes down to this: science charts the movements of things in space and time. What agency compels them to move in this fashion it never, never, never ascertains. That conception of agency is a figment of folk-lore and can never enter into scientific analysis in even the slightest degree. This means that from the human point of view science is utterly unsatisfactory. But especially it means that science never solves the mysteries of the Forces which rule our lives. In the field of mystery—and human life is all mystery—religion remains supreme.

For putting science in its place, however, this argument is still incomplete; so it has long been a commonplace of the professors of philosophy that Hume's work required to be completed by Immanuel Kant. Both Hume and Kant were versatile men, no less than the Walrus and the Carpenter, and talked of many things; and it is a little rash, perhaps, to speak of either as though his one mission in life had been to make this particular point under discussion. As a matter of strict historical fact, nothing of the sort is true. But almost any figure in history would be amazed to discover what he has become famous for, and if he chanced to be a reflective man he would be forced to reflect that in immortalizing him in this fashion posterity had only set up an idol to its own prejudices to which it ventures to believe that he contributed support. So it is with these philosophers. Hume wrote a history of England, and Kant discovered the nebular hypothesis. But what each is chiefly remembered for, and thereby selectively immortalized in the history of philosophy, is Hume's dissection of the category of causation and Kant's completed exposure of the inconsistency and "unreality" of the world described by science. It makes no difference that neither Hume nor Kant took the least stock in revealed religion, and that two more thoroughgoing "agnostics" would be difficult to produce in any period of history. A man may make a fortune

SCIENCE: THE FALSE MESSIAH

manufacturing the munitions of war and spend it for the endowment of efforts to preserve the peace. His place in history will then depend upon the turn of subsequent events, after his death, over which he has no control. The events which have placed Hume and Kant are the episodes in the struggle of civilization to assimilate science. They constitute one, or two, distinct episodes, and this defines their "work."

Kant's motive in this argument was to exhibit the frailty of the human intellect. With this end in view, he took a group of the most commonplace notions and carried each to its ultimate conclusion in two opposite and contradictory directions. For instance, time. Now time, he said, must have a beginning. For if it had no beginning, then an infinity of time must have elapsed already at this moment. But that is inconceivable, since it is the nature of eternity never to be over. Therefore it is the nature of time to have started at some remote but finite initial moment in the long ago. But, on the contrary, if we conceive time to have started at such a moment, we have to imagine it to have come into being after a previous condition of things in which no time was. This, however, is to imagine a time before time, which is an absurdity. Therefore it is the nature of time to have been running always, and a beginning of time is inconceivable. Thus time combines, in the catholicity of its nature, two squarely contradictory notions!

Faced with such a contradiction, a sensible man

would turn tail, not on his clock but on infinities, on the ground that he has some use for time but none for either pre-existence or eternity, and this is precisely what Kant advises him to do. With the calmness of a man who is not in the least taken in by his own trick Kant advises his readers that this demonstration has no effect whatever for all ordinary uses of time, or any of the other notions which prove similarly contradictory. It does not even affect the employment of infinity as a mathematical symbol in calculations involving time. That infinity is an imaginary number, like zero, a process of figuring, not a condition of supernatural existence; and such calculations manipulate cardinal numbers symbolizing time, not "time itself," as "pure duration" or any other sort of philosophical "condition of existence" in this world or the next. The Kantian demonstration shows that every conception or formula has a limited usefulness and can be warped quite out of shape if subjected to tests for which it was not designed, especially the test of Eternal Truth. But this in no way affects the usefulness of finite clocks, nor the truth of finite time. Thus the world of clocks and time is made safe for science, while Eternal Truth is forever preserved from contamination by such mundane conceptions.

This segregation was greatly to the advantage of science. Perhaps indeed that alone made science and the new age possible. Whatever we may think now of its subtleties, metaphysical dualism gained for science the pontifical absolution and the free-

SCIENCE: THE FALSE MESSIAH

dom to go and sin some more, an advantage which science has certainly exploited to the fullest possible extent. And the pontifical blessing was bestowed only after the philosophers had demonstrated that science can never, by any conceivable discovery or hypothesis, disarrange Eternal Truth. The dualism was also, obviously, of equal service to Eternal Truth, which it thereby guaranteed. The waters of science might rise. They might cover the earth. The immortal soul, with all its sacred preoccupations, was safe in another universe. Secure in the bosom of the Almighty, it could now survey the scene of time and tide with serene indifference, turning unseeing eyes upon the Great Illusion. Science might be supreme in the realm of things as they seem to be; in the realm of things as they Really Are, the folk-lore—or at least philosophy—remained impregnable.

Such is the modern dualism. No other people has had one exactly like it,—since no other people has been exposed to the hazards of science and technology. Our civilization is itself a dualism of technology and folk-lore. Bodies and spirits have abounded in other periods of course. The seen has been supplemented in imagination with the unseen. But we must remember that the seen is always no less a work of the imagination than the unseen, and where the two consort harmoniously together as agents that differ only in relative potency, they form not only a harmony but a working unity. It never occurred to the early Christians—certainly not the early Hebrews—to split their universe into a duality of men and angels: the Lord of Hosts (spiritual) and His earthly children (material). It is of no particular advantage to a Zulu to differentiate between man the farmer and man the devotee. Among Zulus, as among most of the peoples of the earth, these two functions are carried on side by side. Husbandry and devotionals are freely intermingled. That has ceased to be the case among us. The technology of husbandry—to say nothing of manufacture—is now so potent that technological devotions have been almost wholly crowded out. It has also become so extensive that no mundane field of operations has been left in which the potencies of religion are still to be relied on.

Thus the unseen agents of the spiritual world have been crowded out of the primordial unity into a newly invented dualism because there is no longer any place for them in human life. If religion had been inseparably attached to a certain area, it must either have disappeared or, at the last stand, have saved itself by wiping out its enemy. If certain operations, such as have been performed not with tools but with incantations and invocations, had continued to be the locus of our folk-lore, then it would be impossible to perform them with tools, and technology, and science, without finally upsetting all religion. If rain-making were always and forever the field for prayer and fasting, a scientific weather bureau would precipitate the death struggle between science and religion. We have safely inaugurated the weather bureau, by an extraordinary feat of vivisection.

For this task our superior sophistication, born of technology is a great—if not indispensable—advantage. When people have no notion whatever of the distinction between respiratory organs and digestive organs, they are less inclined than we to draw a sharp line between body and soul. We are accustomed to anatomical dissection. Consequently we have no trouble conceiving the division of body and soul along anatomical lines. The soul, we think, is a sort of vermiform appendix. It is of no practical importance in any of our affairs. It is an interesting survival from a period of earlier usefulness. Therefore it is sacred, always and forever a part of the Complete Man, a symbol of the eternity of Life, to which a man's days are but a candle-flicker. We may separate it from the active tissues of the body by the surgery of dualism; but we preserve it in a state of leisured dignity, immersed in holy water, and enclosed in an unused cupboard known as the Sabbath Day.

CHAPTER XIII

PHILOSOPHY EMBALMS A FOLK-LORE

THE intellectual defense of the "advanced monotheism" of western civilization finally comes down to this: it has become necessary for us to understand and demonstrate the infinite powers of Almighty God. What an achievement—to bring Omnipotence within the compass of the human intellect! To verify Divinity by an exercise of logic! Yet as an exercise of logic it is not so difficult. Such an expression as "omnipotence" is used wherever gods obtain. The god, or, in cases where there are several, some one among the gods, is invariably pictured as omnipotent, and the meaning which has clustered about that word in the imagination of unphilosophical people is simply that of a god who can produce any effect he sets his mind to, a god mighty in battle, a god strong and able to prevail, a god who is more than a match for the neighbors' gods and of course more than a match for all the other divinities of his own domain, both his own camp-followers and the bad characters and outlaw deities of his spiritual habitat. Such a god is omnipotent in the sense that he can prevail over all others, a natural and sensible conception found in the folk-lore of nearly all the peoples of the earth.

SCIENCE: THE FALSE MESSIAH

Modern omnipotence is a different matter altogether. If we reason it out, omnipotence can not mean "potent over all" for the Infinite for which there are no others. What can it mean? It can mean only the powers of the Being in relation to whom there are no others, the primacy of a champion who can meet no challengers.

A strange sort of omnipotence? From the human point of view, yes; but modern theology is nothing if not strange. The familiar sort of omnipotence, sensible enough as the rating of the senior spiritual officer, can not stand within the jurisdiction of science. According to science no single force exists in the universe capable of prevailing over other forces. Indeed, what is the universe of science but a balance of correlated forces? This is a world of machines. But it is also—now-a-days—a world of men. Therefore if the omnipotent one is to issue a real challenge and attract actual opponents, he must do so in this world of actual events and moving forces, the world in which successes and defeats are scored, and the world in which nothing is supreme. The only opening for omnipotence, therefore, is to retreat from this world altogether to a region where positive forces are not checked by negative, and gravity by centrifuge, a region where irresistible forces and indestructible barriers can exist together in one and the same Being.

In short, infinity and eternity are no longer the laudatory titles of the tribal god. When applied to the one true God of "advanced monotheism," they

PHILOSOPHY EMBALMS A FOLK-LORE

do not mean unlimited jurisdiction actually enjoyed by a sovereign whose boundaries take in all known regions, nor the uncounted years of a creator who remembers when the world was young, has lived through all the vicissitudes of human decline and regeneration, and will definitely survive to wind up the business when the last books have been balanced. Such naïveties have been blue-penciled by our scientific education. These sophisticated infinities and eternities are more precisely described as timelessness and spacelessness. Here we have truly metaphysical attributes of divinity. The universal Being is infinite in the sense of the disappearance of all degrees of magnitude, and eternal in the sense of a void in which nothing happens: neither the ticking of a clock nor the succession of the seasons. All is change—where there is no perceptible difference between changing and standing still. Naught but the permanent changes—where permanence is the lot of a Being that neither comes into existence nor passes away. In short, the Absolute (as God is most appropriately called in metaphysics) is that order of reality in which whatever can be conceived to exist or happen is imagined not to exist or happen. Such things can be imagined by the same technique with which we imagine the square root of minus two—by formula.

The spiritual nature of the Absolute is similarly esoteric. It is easy to credit that such higher powers and magnitudes must be immaterial. The difficulty lies in fabricating such an immateriality as

SCIENCE: THE FALSE MESSIAH

will be appropriate to so extensive an infinity. The commonplace spiritual elements of which the immaterial beings of folk-lore have been composed will hardly do. The Absolute is not immaterial as elves or angels are immaterial, as Jahve and Beelzebub are immaterial, or the human soul in purgatory. Elves dance and angels sing; Jahve and Beelzebub carry on interminable internecine warfare; the souls of human beings writhe in torment. The Absolute is incapable of these experiences. But the Absolute is equally remote and superior to erosion by the elements of which the material universe is composed. It must be non-material: It can not be spiritual in any ordinary sense. By the same formula which succeeds with other attributes, It can be clothed in the opposite of materiality. It can be immaterial in precisely the sense in which It is infinite, a higher immateriality than that of the merely invisible, a ghost in the closet or an image in one's head, a higher spirituality than the gaseous residue which is popularly supposed to be released from a dying body. The Absolute is Pure Spirit. Whatever the Absolute is, It is always pure.

This purity, indeed, is somewhat excessive. The effect of it is to make the Absolute unapproachable, a very genuine misfortune, since the human motive behind all this speculation is the common human interest in a savior of humanity. And that the Absolute, in its capacity of universal negative, can never be. Consequently many thinkers have been led to assign to the Absolute the qualities of the human mind, to represent it as a higher order of intellect and purer distillate of soul. According to this doctrine the Absolute is "of the nature of" the human spirit, is the same thing, indeed, clarified, and sanctified and thus relieved of all the traces of its bemeaning association with the body upon this earth. But this, as every one can see, is a false doctrine, a sort of metaphysical heresy; for if it is a heresy against Roman Catholic theology to reject the trinity of the godhead in favor of the Unitarian view of the matter, it is equally a heresy against the Absolute to reject strict unitarianism for a trinity of mortal mind, immortal soul, and Absolute Spirit. Conceived in such terms, the Absolute Spirit immediately ceases to be pure and undefined and becomes an obvious amplification of the tribal God of the European peoples.

To avoid such heresies, the German metaphysician, Hegel, devised a formula, a sort of patent Absolute detector, the famous Hegelian dialectic. This machine works in the following manner: If ordinary thinking suggests that there are two sides to a question the Hegelian dialectic denies them both and unites them both in a "synthesis" which is neither and both at the same time. A synthesis of this kind avoids all the difficulties of the ordinary manner of thinking in which if it isn't light it must be dark. Thus the higher synthesis of the issue between the loneliness of a Being that is infinite and the finitude of a Being that enjoys company is a state of one-and-many, peculiar to metaphysics, in which all these limitations are impartially denied while at the same time all are collectively claimed as the powers of a Being whose especial prerogative it is to be opposite things at the same time. The freezing and boiling points of Pure Spirit are determined by just such a synthesis. Since it is neither matter nor mind of any perceptible or intelligible sort, the Absolute is a still purer substance, partaking of the qualities of both. That is absurd? Certainly, but no more absurd than any other Hegelian synthesis. The physical world, it must be admitted, is known to us only through the mind. Science itself is a knowing. Meanwhile the mind is known to us only in action, through our perception of the physical world and the conception we form about it. Thus it would appear that neither is complete, neither is the whole of reality, each is an aspect, and there is contained even in these words the suggestion of a higher knowing: "We know" each aspect through the other; a knowing that is above them both, and of which both are expressions. This is pure knowing. The Absolute is such a Complete Knower. In the Infinite Mind of the Absolute, body and soul are one; the dual aspects are conjoined; unity prevails. The Hegelian transformation is complete.

These demonstrations leave an unavoidable impression of verbal trickery. In particular, the technique by which a higher order of knowing is suggested in such an illustration as the foregoing looks decidedly like a grammatical imposition. An unfair advantage has been taken of the looseness of the vernacular, almost as though some one were to argue the existence of Jupiter Pluvius from the presence of the pronoun in the expression, "It rains." According to the usage of the common tongue it is possible to say that "we know" matter through the mind, and also "know" the mind through our conception of matter. Indeed, one can say such a thing in no other way. But whether these tortuosities of the language prove anything or not is hardly open to argument. Of course they prove nothing whatever.

In a general way, however, the principles of grammar do show that the permutations and combinations of words are not limited in the least by the meaning of the words. All that is necessary in order to compose a sentence is to set down a subject, a copula, and a predicate. All that is necessary to produce a negative is to add to any word the prefix of negation. All that is necessary to achieve infinity is to draw out any quality to the superlative degree. These operations can be performed strictly according to the laws of grammar, and will thereafter make linguistic sense and so will certainly seem to mean something whether there is anything in heaven or earth, Horatio, that our philosophy has ever dreamed of, or only the absence of any advices to the contrary, a shortage of information, a vast unknowable—behind the beyond, as Mr. Leacock aptly puts it.

Now it is the avowed object of metaphysics to

SCIENCE: THE FALSE MESSIAH

penetrate behind the beyond, and that is by definition a region to which the mind of man can never penetrate. Consequently, the sum total of all our information about the Absolute adds up to nothing, net. The most penetrating philosophers, the leaders of the clan, the men to whom every one does honor as the most daring of all the cosmic explorers, have invariably brought back this report. It is the lesser men, the men whose humanity has got the better of their dialectics, who have seen across the wastes of metaphysics the shining mirage of human aspiration and have come back with visions of the New Jerusalem, an Absolute created in the image of an elderly professor, a sort of dean of the universe. They are the metaphysical heretics. But if we take their arguments, their actual analysis, we find that the mirage is in the sky, while the city on the ground is as dusty and smelly as before. For example, a great American writer of meta-heretical temperament has preached most movingly on the theme of "loyalty to loyalty." The face value of his argument is that any particular manifestation of loyalty is a finite instance of a quality which in its highest and purest form is superlatively good. In being loyal, therefore, we are partaking of the infinite good, we are achieving a fragment of perfection, fulfilling in tiny part the purpose of the Infinite. This makes loyalty quite alluring. But the metaphysician's argument contains more subtleties. No doubt it is true that the Absolute is the exponent of the highest Loyalty. However, the other qualities similarly exemplified in the highest degree give the Loyalty of the Absolute a significance quite out of relation to the adherence of mere crawling mortals to the object of their desires, so that it may be true, metaphysically speaking, that each finite loyalty is a tribute to the Infinite and still be equally true that in human terms every human attachment is good or the reverse according to what the object of the attachment is. "Loyalty to Loyalty" has a heart appeal far beyond what is exerted by the parts of the verb to be in which Hegel usually worked; but it has its dialectical flaw. All these expressions are undeniable and meaningless.

In the end, the conclusion is inescapable that this is the object of the metaphysical reasoning: to be undeniable and meaningless. What God has it wrought? One who is demonstrable and powerless. The tribal gods of simple peoples are not intellectually demonstrable. The members of the tribe believe in them and believe in their good works, or bad, as the case may be. But that is faith. Their only argument is that every one believes. The Holy Church pronounces so and so, and the church is of course infallible. This is not proof. All the beliefs in the actual potency of the ancient Christian God went flat under the steam roller of scientific demonstration. The God of philosophic speculation is indeed beyond the reach of the scientific steam roller. But by the same token His children are beyond His reach—if it is proper to assign the parental rôle to the Absolute, who must be, in

SCIENCE: THE FALSE MESSIAH

Hegelian dialect, father and children both, the Absolute Parent, a level of reality in which families are united and there is no marrying or giving in marriage. The Absolute is the God not of love but of Being. The God of metaphysics ineffably exists, but is nothing-in-particular, is nothing that anybody wants.

The object of these ratiocinations, be it always fondly recollected, is a true and just conception of God—a folk-lore worthy of scientific enlightenment. As professional thinkers, the philosophers will no doubt place heavy emphasis upon the truth of their conclusions. They will even insist that theirs is a responsibility for coming at the truth though the heavens fall. No doubt it is. Scientists also make a habit of insisting upon an interest in nothing but the truth, whole or not. Theirs not to reason why, theirs but to verify. Yet it remains as true as anything can be in this not very intelligible world that however pure may be the motives of an individual scientist of anything resembling pecuniary interest, the purpose of science as a whole is to be applied. If the truth thus resolutely wrested from an intractable cosmos is not to be applied to the glory of man, it is utterly pointless and might as well remain unknown.

In a similar fashion the truth of metaphysics has no other object than the glory of God. Let it be as true as may be; if it does not so redound to the glory of the Most High—or, failing, turn in a frankly negative report—it is similarly pointless and had much better be abandoned as a scandalous waste of time and intellectual energy. There are no two ways about it. Knowledge for knowledge's sake is all very well as a motto to preserve the innocence of individual investigators; for knowledge as a whole and for civilization the phrase is meaningless. In the history of civilization, these speculations represent the attempt of the modern intellect to square theology with science, and they can be judged in no other way. The scientists who spend their declining years puzzling over the "significance" of the "basic" conceptions of science are not merely interested in the limitations of the scientific method. They are looking for significance of another sort, significance for "the relation of man to the universe," in other words, religious significance. The philosophers who take their broadest generalizations and work them into esoteric and unintelligible patterns are not playing hide and seek with words in a game of competitive unintelligibility. They are seeking that they may find. It is just as important therefore to inquire whether their findings are intelligible or reliable. No doubt the Absolute is reliable, whether intelligible or not. Those who claim to understand It maintain their position with flawless logic against all comers. These several distinctions of substance and relation, existence and subsistence, and so on, these durations and relativities upon which the most cautious contemporary thinkers dote, are all logically sustainable as meaning something or other. But there is also the question whether what they mean amounts to anything. All of which sums up to this: whether an intellectual abstraction of any variety is a satisfactory substitute for God.

This is not an intellectual issue at all. It concerns, on the contrary, the common human interests of common people who have done business heretofore upon the basis of a folk-lore and have no other basis upon which to proceed. Their question is: Can any vestige of that folk-lore be maintained? The God of their religion has never been an intellectual abstraction. He has been, if we may say so, a human God, even a humane God. Certainly He has belonged to the working classes. That is the whole reason for their interest in Him. They have been convinced, rightly or wrongly—and which we think it is matters very little, as history has amply exhibited—that the God of their fathers has taken a personal interest in them, that He has been in some sense accessible to them, that He will save them at the last. The manner of the saving has no doubt varied greatly as well as the details of His accessibility. But behind all variations the fact of solicitude and potency has never flickered. Wherever argument has entered in it has affected only the manner of salvation. The fact of salvation is a fixture, indubitable as long as a religion stands. When and if it fails, that religion is done for. Folk-lore is not possible upon the basis of intellectual abstractions. The "purer" the abstractions the greater their failure.

PHILOSOPHY EMBALMS A FOLK-LORE

Now if one thing is clearer than another in metaphysics, it is that in every field and department, in every school and philosophical tradition, modern metaphysics has turned out nothing but intellectual abstractions. What human potency has this Pure Being? None whatever. For the remission of sins, for guidance upon earth and salvation hereafter, it is as powerless as a Christmas tree locomotive. God, freedom, and immortality, we are told, are the three great conundrums of theology and metaphysics. What does modern philosophy teach concerning these things? Every school has solved all of these conundrums. No doctor of philosophy who could not undertake to unhorse all three in fair encounter would venture to accept an instructorship in any university. But are any of the variously demonstrated Most Perfect Beings, or failing that, Ultimate Realities, in any way at all accessible to man? None whatever. The subtle gossamers of definition by which it is shown that the human will is wholly free without being one whit the less subject to all the winds of causal compulsion that blow through all the universe—does any one of them bring conviction of sin and the means of salvation? Hardly; such expectations are now-a-days quite unphilosophical! Does any intellectual theory now in good standing extend the slightest hope of eternal bliss for repentant sinners and punishment hereafter for the damned? None! From the point of view of Truth, modern philosophy is no doubt unexampled. Under the direction of science every superfluous ounce of superstition has been eliminated. Is it therefore beautiful? Has it, then, found any truth which any one desires in the least?

There can be no two answers to this question. Metaphysics is the sarcophagus of the spiritual life. Within it our ancient and cherished folk-lore lies embalmed. The art is perfect. Its product is a mummy, swathed in rich symbolic trappings, but quite dead.

CHAPTER XIV

THE REFORM OF SUPERSTITION

RELIGIOUS tolerance is one of the principal ideals of modern European culture. It is a peculiarly American ideal, perhaps because America is the revised version of Europe. But whither does it lead? According to Mrs. Hemans, our fore-fathers suffered the hardships of the *Mayflower* and the rigors of a stern and rock-bound coast all in the interest of "freedom to worship God." Just what they, as distinct from Mrs. Hemans, meant by this was a little obscured at the time by their attitude toward Roger Williams; but let that pass. The strange thing about the migration of the Pilgrims is that they apparently had no idea of opening up a new continent and setting forth upon it a new nation. If that is so, the history of their descendants is indeed remarkable. What they sought was the privilege of reading their Bibles in a certain way; and what they got is (for the moment at least) economic world dominance. We must infer that all these things were added unto them, a veritable shower of blessings!

The case becomes all the stranger when we reflect upon the fate of those theological niceties for

SCIENCE: THE FALSE MESSIAH

which the Pilgrim Fathers made the supreme sacrifice. The austere "Congregationalists" of sixteen twenty would be rather dismayed to find that they have become the sign and symbol of the highest caste of Oldest Inhabitants and that their *Mayflower* hardships have been immortalized in the luxurious appointments of the presidential yacht. But they would be still more disturbed by the realization that—the country having at length been made safe for Baptists, and Quakers, and other peculiar people—those very sectarian distinctions for the preservation of which they so nobly struggled are about to be abandoned in the interest of a wider, fuller, and freer Christianity. It would seem now that the principal abuse to which Christianity is subject is the abuse of sectarianism. Here we have a score or two of Christian denominations, each maintaining its churches and its clergy, its foreign missions and its domestic charities, and conducting them all with strict reference to peculiarities of doctrine and ceremony which serve only to set the fundamentals of Christianity at nought. All these churches are founded upon the saving power of the love of God. They all preach Christ and Him crucified. The differences of interpretation and of forms of worship are all relatively superficial, and are becoming more and more so as all the various interpretations become more and more liberal. Congregationalists still sprinkle and Baptists still immerse. Quakers still wear their hats in church and segregate their women. But they do all these things with much less zest than formerly. The conviction is gaining ground that they have become abuses, at least in their strict application, and can only hinder the development of true Christianity by dividing Christians against themselves. Nonsectarian community churches are multiplying, and inter-denominational committees are becoming daily more important in the councils of the clergy. There is no doubt about it; the rugged tenets of the Pilgrims are about to go.

The trouble with the Reformation, the trouble with all reformations, is that they never know when to stop. This is because they are essentially negative. The process can never stop, short of total extinction, because it has no definite, foreknown objective. It proceeds by reducing. The reductions are all abuses. The argument of the inter-denominationalists is perfectly sound, granted the present situation. It is true that no one of these sects is inclined to-day to rest its case for salvation upon the uniqueness of its sectarian doctrine. Congregationalists do not preach the damnation of Baptists any more, or Baptists the damnation of Friends; and since this is so, there is no compelling reason for any denomination of Protestants holding aloof from the brethren of the other denominations. Every decent modern Presbyterian feels that it were better that a man should be brought to God though he become a Methodist than for him to remain forever outside the fold of "the Christian church." Once this is admitted, inter-denominationalism is inevitable. The disabilities of divided councils far outweigh any mere preference in the forms of worship.

But this assumes modernism as a preceding condition; whereas modernism itself is of recent growth, a late product of the Reformation. It came as a protest within each denomination, against the abuse of strict interpretation. The disagreeable character of this abuse, also, is incontestable—assuming the situation; and that situation is one from which the Protestant Church could hardly have escaped. The essence of Protestantism, as contrasted with Catholicism, is the "freedom of the individual conscience." In the various sects, and at different periods, single individuals have enjoyed sometimes more freedom and sometimes less. Even to-day, for example, neither Presbyterians nor the most latitudinarian of low church Episcopalians set any great store by the doctrinal inspirations of individual laymen, or for that matter individual clergymen. Nevertheless the principle is there. All these churches accepted the principle of free interpretation in some form or other as the issue between them and the Church of Rome. At the very least, they rejected the authority of the Pope and the Holy See. They are committed to using their own judgment in the interpretation of divine revelation.

Moreover, to confuse their case still further, each Protestant denomination faced this issue and rejected the authority of Rome on some particular point of ritual or doctrine. To all Protestants,

THE REFORM OF SUPERSTITION

"popery" means not only the intolerable authority of a foreign autocrat in matters of individual religious conscience; it also means a certain particularly objectionable abomination: the polytheistic cult of the Virgin, the idolatrous worship of the saints, the use of icons, incense, altars, or holy water. The feeling among Protestants is that such items of belief and practise—confession, or absolution, or transubstantiation—are medieval, barbarous, and altogether unworthy the representatives of an enlightened civilization whose sole object of worship should be Omnipotence and whose sole mode of worship should be the sincere opening of the heart to the love and mercy of Almighty God.

Having assumed this position in some degree or other, the Protestant churches have therefore been peculiarly at the mercy of what we call modern progress. They may have rejected the miracle of transubstantiation because it seemed unreasonable to suppose that the holy wafer can be transmuted into the actual flesh of Christ by the ministrations of a "dirty priest," as they not infrequently put it; and they may have accepted the miracle of the transformation of water into wine at the marriage feast at Cana because it seemed reasonable to believe that such a thing could be accomplished by the Son of God. But having once accepted such a test they were once and for all upon different ground from the "Universal Church." As time passed, and more became known, or scientifically believed, concerning the constitution of liquids such as wine and

SCIENCE: THE FALSE MESSIAH

water, the possibility of water changing into wine became more and more difficult—upon the basis of reasonability. To be reasonable, one must believe in science. To be theologically reasonable, one must true the miracles with science.

Hence there arose all the long agony of adaptation to which Protestant Christianity has been subjected, the agony of adaptation which produced Modernism. Having expunged the abuse of ecclesiastical authority, the church found itself without an anchor, and able to maintain its equilibrium only by riding before the wind. It was carried into the latitude of Modernism as a consequence, and without being able to stop there. There was no possibility at any time of coming to rest upon the doctrines of that time, because the principle of Reformation, on which the church was founded, is a principle of disequilibrium and instability, a principle of negation and continual, progressive expurgation of successive "abuses." The Catholic Church may or may not have had similar experiences. It probably has had; its inner history is not wholly unlike its "external" history. But the Christianity of reformation, the distinctively "modern" Christianity, is a product not of growth and enrichment of belief and worship but of dilution, filtration, and distillation.

If we imagine this process carried through to a complete inter-denominationalism, we are no doubt confronted by a new "universal" church. But it is not the Universal Church of old. Indeed, the differ-

THE REFORM OF SUPERSTITION

ence is most informative. The essence of the Universal Church was power, secular and ecclesiastical, based on centralized control. This power may have been external to religion, strictly interpreted as an adventure of the soul. Let us suppose it was. Let us suppose that it rested on the centralization of all medieval society by a feudal system of which the Universal Church was an appropriate part. This extraneous element of feudal authority is what the Reformed Church broke away from. Thus we may suppose that the Reformed Church carried with it the essence of religion, the essence of Christianity, leaving behind only the medieval social system. But the interesting circumstance is that having done so, and having come through, as we may imagine, to full maturity in the new Inter-denominational Union of Reformed Christianity, the modern church is not only quite clear of feudalism; it is wholly without power of any kind. Indeed, the more inter-denominational the universal Christianity becomes, the less power it is going to have.

The power of the Church of Rome, a thousand years ago, rested upon this circumstance: that it could excommunicate and thereby effectively expel a man from every church in Christendom. It could expel him from the whole community of respectable Europeans. Inter-denominationalism is the reverse of this. Its whole objective is to make Presbyterian and Methodist churches interchangeable; so that the man who has been at home in one will be at home in the other; so that each may do the work of

SCIENCE: THE FALSE MESSIAH

Christianity for its neighborhood without competition or confusion. The whole effect of this is to make it easier than ever before for any man to find a spiritual haven. If he does not like the ways of one, he can try another with no sense of spiritual risk. There is no thought of excommunication here. The object of the merger is to open doors, not to close them; to decentralize still further the worship of God by the guidance of the individual conscience. Thus the inter-denominational church adopts the form of universality, to be sure, but it does so in token not of power but of the complete unprecedented surrender of power.

This may be religion in all its purity as an adventure of the soul. But as an adventure of the church it is unique, and no one can pretend that the church is likely to become a moving force in the modern world as a result of it. Whatever the pure form of spiritual experience may be, human experiences are never pure. Institutions always interlock, churches no less than others; and a church which interlocks with nothing else in civilization but exists only as a sort of Elysian field for pure spiritual experience is a pure social vestige, as ineffective as the vermiform appendix. The purity with which the modern church is blessed could never, credibly, have been accepted by any church of its own free will, for reasons of its own creation, and for the better expression of its own inner purposes; and as a matter of fact it was not in this case. Caught between two worlds, coerced by two opposing forces—

THE REFORM OF SUPERSTITION

one the force of authority, expressing a civilization of feudal hierarchy; the other the force of individuality, or reason, expressing modern civilization whatever it may be—the Christian church has chosen to be reasonable and modern.

The question remains whether any appeal of reason to modern civilization can be effective. Has the modern world as yet developed a civilization?— a positive scheme of institutions and ideas, a stable power-unit centered in science or democracy or something of the kind? Or do we live only in a process of transition, itself a progressive dilution and distillation of ancient European civilization, making way for another as yet unrealized order of which the only positive hint thus far is machine technology? Is modernism what it is because civilization is what it is, the blind leading the blind? The actual substance of our modernized religion should throw some light on these questions. Most of us feel inordinately proud of our reasonable pieties. But this may be a delusion on our part. We may be proud of a laborious achievement because of the effort it has cost. It still remains to show that the result was worth the effort.

Strangely enough, none of us is precisely satisfied with our achievements in this field. We are the prey of enervating doubts. Fundamentalists— worse luck!—are not beset with doubts. They are as liable to quarrel as the rest of human kind; but however much they quarrel, there still remain to them an established church and a consecrated dogma. The church pronounces the dogmas, and the dogmas stabilize the church. But modernists discover to their sorrow that a reformed church weakens the authority of its own creed, while a renovated creed saps the foundations of its own ecclesiastical establishment. The contemporary leaders of reform do not appear to be aware of this dilemma. Some are caught upon one horn, some on the other. Theologians, like for example Professor Kirsopp Lake of Harvard, are extremely sensitive to intellectual difficulties. Churchmen, like for example Professor Harry Emerson Fosdick of Union, are correspondingly sensitive to practical problems. The intellectuals ask, "How can the church pretend any longer to depend upon such attenuated beliefs as even reasonably well-informed people can now hold?" And as if giving the antiphony of a chant, the practical men respond, "How can we make such attenuated beliefs the basis of the church's appeal?" Reformed superstitions are no beliefs at all. A dis-established church is no church at all. When both occur together, how can they appeal to each other for relief? Enlightened unbelief is not a very powerful slogan for the promotion of a church, and a church membership that is wholly optional and voluntary is not a potent buttress to belief. Such is the dilemma of the modernists.

This dilemma, in which the reform of our religion has come to its uneasy end, is characteristic of modern civilization. We have undertaken to become at the same time enlightened and free. Confusion is the inevitable result. It is a hopeless confusion because the whole process of reform is negative. No doubt something will take the place of the pieties we have reformed. But it will be another piety, a belief in something else, a respect for authority of another kind.

CHAPTER XV

SCIENCE BETRAYS RELIGION

THE performance of the scientists who have arisen in recent years as the defenders of religion is indeed an amazing one. But more amazing still is the state of mind of these self-appointed defenders of the faith. Not only do they assume infallibility as a matter of course, but they seem to imagine also that all the rest of the population is so simple-minded that no one will ever see through the circumlocutions and transparencies in which their ambiguous oracles are couched. Apparently, they have made up their minds not only that scientists can say no wrong but that they can talk nonsense indefinitely without detection. For ways that are dark and for tricks that are vain these modern grandees are peculiar!

Scientists are not all alike, of course. There have been scientists who felt the "warfare of science and theology" most keenly—felt themselves, in fact, to be the natural enemies of theologians. As a matter of temperament, they assumed hostility to revealed religion, and declared themselves infidels, or agnostics, or monists, or mechanists, "on principle." Thus the great champions of evolution,

SCIENCE BETRAYS RELIGION

Huxley and Haeckel, rather gloried in their heresy, not because biological evolution disproved any one particular revelation of the origin of things, but because the scientific point of view as they thought precluded any sort of revelation. Natural knowledge and religious authority were direct contraries in the mind of Huxley. To him, scepticism was a scientific duty. "It can not be otherwise," he said, "for every great advancement in natural knowledge has involved the absolute rejection of authority, the cherishing of the keenest scepticism, and the annihilation of the spirit of blind faith."

How much this attitude is a matter of temperament is indicated by the case of the "father of evolution" himself. Darwin was not as evangelical as some of his modern successors, to be sure; but he was far from being a fiery heretic. In various passages, he expressed his belief that evolution was quite compatible with divinity, and might even furnish a divine commentary. He would not have said that science furnishes a conception of God; but he did appear to think that evolution reveals God's plan. We are therefore compelled to conclude that the study of evolution does not of itself predispose a scientist either to believe or disbelieve, since both states of mind are found among equally thoroughgoing advocates of evolution.

The will to doubt seems always to claim its isolated victims. Indeed, the more sceptical a man becomes, the more isolated he seems to be. Perhaps there is a generalization here. Perhaps the sus-

SCIENCE: THE FALSE MESSIAH

ceptibility to doubts is produced in men by some sort of social isolation, or at least by their sense of social isolation. They feel cut off from their civilization. Consequently they can venture not to believe in it. Thus behind temperament there lies a deeper circumstance. Huxley does not now appear to have been an isolated man. Yet we have many evidences in his own hand that he did feel cut off from his generation, particularly in his youth. Often in his letters he complained of the repellent contentiousness of his fellow scientists, of his sense of separation from them, of the "impossibility" of a young man's making his living in the pursuit of science. In our day, Jacques Loeb was a fiercer doubter than Huxley, a more obstinate monist than Haeckel. He was also probably the most distinguished "American" biologist. But he was a Jew, an expatriated foreigner, an inveterate individualist in his conception of scientific research, and altogether a lone figure even as a scientist. In the battles of science, he had won medals galore, but he was not a major-general. He was a "character."

If scepticism is temperamental among scientists, so—in all fairness we must allow—is extreme soft-headed pietism. There have been scientists in both ancient and modern times whose zeal for superstition has quite overbalanced their researches. But they are simply dotards, natural or premature. Because an occasional Oliver Lodge has gone from ether to ectoplasm in his eagerness to get into touch with our legendary future life, it does not follow that science is responsible. Of all the believers in raps, taps, and table-tippings, only a very few are scientists. Therefore it seems unlikely that science is the source of the notions of those few. Science may furnish their séances with a lively vocabulary of strictly modern words. But science did not supply them with dead relatives.

In contrast to these extremes, the common run of scientists are neither more nor less pious than the common run of men in the upper middle classes. Scientists were all children before they were scientists; most of them have married while yet scientifically immature, and given all the usual hostages to polite society. This means that they have the same background of folkways and folklore as everybody else. They have the usual stake in civilization. After reaching the age of discretion they may have some doubts about the legendary sanctities. Most men do who are intelligent enough to amount to very much in any profession. But most men keep their doubts to themselves; and scientists have the same motive for doing so that affects their sisters and their cousins and their aunts. The obvious, natural, and human response of any group of assorted scientists to the "warfare of science and theology" is to deplore it, and if possible to ignore it. Such warfare, if it became open and undeniable, would separate them from their families and their childhood friends and almost wrench their own private personalities in twain.

SCIENCE: THE FALSE MESSIAH

Consequently, when the opposition of science and religion is openly bruited as it was recently and notoriously in Dayton, Tennessee, the impulse of all right-minded scientists is to deny the schism, to assure every one, with tears in their eyes, that science is a noble, beautiful, and sacred expression of man's highest aspirations, and that no one could possibly believe more devoutly in God, morals, and Christianity than the research scientist, blessed as he is with a unique opportunity for studying the divine plan in all its transcendently spiritual details. The discord of the Origin of Species and the Book of Genesis is all a dreadful misunderstanding! This, certainly, was the battle-cry of science—if such it can be called—in the Dayton skirmish. To Doctor Henry Fairfield Osborn, president of the American Museum of Natural History and chief of the defenders of the scientific faith, the whole meaning of evolution is the marvelous benignity of an order of nature in which scientific truth is so perfectly adapted to the requirements of religion and morality. To exhibit this marvel, he even pressed the notorious William Jennings Bryan into service. For years Bryan posed the issue thus: "The real question is, did God use evolution as His plan?" Doctor Osborn has reiterated that question again and again. For him it is the whole crux of the matter. (An unintended tribute to its disreputable author!) But the answer of the scientist is, Yes. God did use evolution as His plan. This is the teaching of science. This is what the exhibits at the American Museum prove. Evolution is indeed a sacred theme. Its moral principle, says Doctor Osborn, is "that nothing can be gained in this world without an effort; the ethical principle inherent in evolution is that only the best has the right to survive; the spiritual principle inherent in evolution is the evidence of beauty, of order, of design in the daily myriad of miracles to which we owe our existence." Proving all this, science is surely innocent of every charge of being a disturber of the peace!

To show that beliefs such as these are the common pieties of ordinary men requires little proof. What is less evident is how a scientist can hold them, not as a matter of temperament but as one of scientific truth. As a matter of temperament, one man can face this issue with the scepticism of an American anthropologist who proposed to say to Tennesseeans: "We do not seek to undermine your faith; but we shall teach you science whether it undermines your faith or not." Meanwhile, another can say with a certain American biologist: "I have been at some pains to make it clear that evolution and religion are strictly compatible." The problem is not the pains, but their outcome. Assuming a temperamental disposition to conform, how can scientists indulge their appetite?

The answer is easy: because science is after all neither proved nor perfect. This is pre-eminently true of evolution. In spite of all the research that has been lavished on the subject since Darwin's time, the detailed operation of evolution remains a

SCIENCE: THE FALSE MESSIAH

mystery. Consequently it is fair to say that evolution, though strongly indicated, is not yet finally established, and eminent scientists have said this during the recent tea-pot tempest—so far does the man out-weigh the scientist in periods of stress and strain. Meanwhile other scientists, Doctor Osborn among them, have ventured the supposition that evolution may be after all finally inexplicable except on the basis of a far-off divine event, or an immanent guidance and propulsion. There is nothing in science to contradict such views—science is silent on final explanations!—except the spirit of science, which is undeniably mechanistic. But what is mere spirit among friends?

Indeed, even the spirit of science has recently been called in for reconsideration. Science received its tone from the physical sciences, partly because they appeared first, but more because they are most obviously derived from instruments of physical measurement, and so naturally would come first and grow fastest. For the same reason they are rather definitely mechanical and "materialistic," and so all science has been considered until recently. The discovery of the atom, for example, has seemed to many thinkers to indicate that all things living and otherwise are composed of these minute but nevertheless material bodies. The inference was natural. Atoms were identified by a chain of investigations in which weighing played an important part, and as might be expected atomic weight produced in nearly every mind the impression of substantiality and

SCIENCE BETRAYS RELIGION

materiality, so that physics and chemistry seemed (for the time) to be the most material of sciences. But this odium has been removed by the discovery of sub-atomic particles. Electronic weight has not such significance as the weight of atoms. Electrons are presumably all of the same size and weight and substance (if any); their weights do not vary with the elements, as is the case with atoms, but only their numbers. Furthermore, the electrons have been discussed chiefly in electrical terms; hence their name. Their behavior appears to be electrical behavior to such a considerable extent that no other behavior has so far been discerned. But electricity—what is that? "Nobody knows!"

Similarly, the starry universe—another stronghold of "materialism" ever since Galileo's telescope—has developed unexpected subtleties. The older astronomers searched the heavens and found no God, however much they may have wanted to. The earth and all that is therein were traced to an event, far-off but no more divine than a tidal eruption on the sun. The milky way was counted, or at least computed, its contents classified, and our own planet recorded as a fly on the chariot-wheel of a lesser star. This looked bad for divinity, or so many people including scientists have felt. But we have been rescued from it by relativity. Time and space, it now appears, are quite relative. The universe is warped. Straight lines are not straight when they are the path of light in a gravitational field. Time depends upon the position of the ob-

SCIENCE: THE FALSE MESSIAH

server, and in brief the whole business of astronomy and celestial mechanics is a matter of point of view. The planets continue in their orbits and the sun rises and sets as usual; but we can now reflect that these things are so only within the limits of our poor powers to add and subtract. None of them is "real."

To go from this confession of ignorance to the positive assurance of divine knowledge is a longer step than most scientists care to take. Most of the nice old gentlemen of science have contented themselves with pointing out that now "materialism" has been eliminated from physics and astronomy, we are free once more to believe in God. A biologist such as J. Arthur Thomson, lecturing on science and religion before Union Theological Seminary, or such a mathematician as Alfred North Whitehead, lecturing on science and civilization before the Lowell Institute, is careful to hold his temperament in check. After all, he is speaking as a scientist; and as a scientist he can only say that these great new modern discoveries do explode dogmatic atheism. They do make it possible to believe in God. Of course they do not make it compulsory. The same lack of finality in science which leaves "materialism" unproved leaves everything else equally unproved. Cautious and eminent scientists allow for this. But still, between the lines, by gesture if not by word, they permit us a glimpse of their own steadfast belief.

They have never doubted. And if some one else

will relieve them of the responsibility of uttering the words, they will sign them. Only a short time ago, forty American scientists and university presidents, many of whom might have avoided such definite commitments by themselves, did sign a quite extraordinary scientific "Credo." This remarkable document was written by an electron-physicist of Nobel dimensions, and issued from Washington like an imperial ukase, with signatures attached, in May, 1923. Speaking thus corporately, the scientists laid aside diffidence and caution and announced in no faltering terms the divine authority of their studies. "It is a sublime conception of God that is furnished by science," said they, "and one wholly consonant with the highest ideals of religion, when it represents Him as revealing Himself through countless ages in the development of the earth as an abode for man and in the age-long inbreathing of life into its constituent matter, culminating in man with his spiritual nature and all his God-like powers."

When men are in such a state of mind as this, it can do no good to challenge their "argument." It can do no good to point out that the spirit of science has not changed. Science has never proved materialism, or any other gospel. It has never stopped men believing in God. It has never been any more final than it is now. The openings in the front line of science, do they mean that ground has been lost? Are we then deprived of our scientific instruments and thrown back upon pre-scientific

SCIENCE: THE FALSE MESSIAH

devices either in our mechanics or in our thought? Not in the least. What has happened may be a revolution, as Whitehead says, but it leaves science precisely where it was before. Newtonian physics is no less useful because Einstein has shown that it does not hold good for distances reckoned in light years and for velocities like that of light. The atomic weights have not been appreciably modified by the "discovery" that the electrons are not material particles—much less the substantiality of mountains and Mahomets. Evolution may be unproved, it may remain always "finally" inexplicable. It may suggest to some minds the probability of an omnipotent propelling force. Nevertheless, the descent of man remains the ascent of ape: not of any existing ape, to be sure—we are not baboons or orang-outangs!—but the ascent of some nicer ape, some sanctified brute, predestined by Omnipotence to be our parent, the undoubted anthropoid!

Neither is anything to be gained by pointing out that "God" is not a scientific conception, that scientific researches furnish no conception, sublime or otherwise, of anything that can not be approached with instruments of precision, that spiritual principles are as irrelevant to biological evolution as Jabberwocky. Every scientist knows this. No one understands better than the Nobel physicist and the president of the American Museum that in the process of biological evolution the one test of fitness is the fact of survival; that in the history of species fitness and survival define each other, so that sur-

SCIENCE BETRAYS RELIGION

vival proves nothing whatever but such fitness as led to that particular survival, and fitness means nothing more than whatever capacity—whether the gigantic size Jurassic reptiles or the spiritual nature and God-like powers of modern physicists—did in fact, for the time being, survive. Nothing could be further from any "spiritual principle" than biological evolution, let alone celestial mechanics or sub-atomic physics. But all this is beside the point. The apologists for modern science do not intend that their words shall be taken literally, or even convey any precise meaning, scientific or theological. They are nice old gentlemen who are disturbed by modern unbelief, and they are making the supreme gesture of conformity. They are offering prayer in public. They are saying, "Look at us. We are eminent scientists; and yet we are superintendents of Sunday schools." To which the only appropriate reply is the vulgar one: "Think of that and bust out cryin'!"

The occasion is indeed one for lamentation. These men are moved to social and "spiritual" conformity not because they are scientists but because they are gentlemen. They are anxious about materialism and unbelief not because they are scientists, or gentlemen, but because they are elderly. Being elderly, they are appalled at the all too evident prospect of great change. Being elderly, they occupy positions of responsibility. It was all very well, we can imagine their saying, for Huxley to refuse to believe on any but scientific grounds

SCIENCE: THE FALSE MESSIAH

"even if the refusal should wreck morality and insure our damnation several times over." The situation was quite different in Huxley's time. For one thing, morality was then in no danger of being wrecked. He lived at the height of Victorian propriety, before the War, before Bolshevism, before civilization arrived at the cross-roads, before the passing of the great race. To-day we are faced with all these things. They are facts. We know now what the wreck of morality might mean. Furthermore, though Huxley made himself the spokesman of biology, he was not the spokesman of society. Then the scientist was an outsider, a knocker at the gate of civilization. To-day he sits in the seats of the mighty. He is the president of great universities, the chairman of semi-governmental councils, the trusted adviser of states, and even of corporations. That indifference to consequences which is possible, if not proper, to a scientist who is only an irresponsible teacher and research worker is not possible to scientists who are the custodians of civilization. When you become a custodian of civilization, the most important thing in the world is civilization, more important even than the science through which you rose to your position of power and esteem. This is the spirit not of science but of ex-scientists.

In a less enlightened age, such a lapse from research to invocation would be less remarkable. Science has not always held the keys of life and death. Primitive scientists did not claim to control nature. They might apply new alloys to the art of plowing or its twin technique of beating plowshares into swords. But the fate of battle was still in the hands of the Lord of Battle, as it had always been. The arts were a small part of concerns much larger than the arts. Prayer was not an accessory to keeping one's gunpowder dry, certainly not a secondary afterthought. On the contrary, the gunpowder was secondary, a gesture of self-help primarily valuable as an invocation to a deity who helps those who help themselves. From the point of view of primitive men, the plowing of fields and even the sowing of seed are by no means the central reality of husbandry. Their festivals of germination no doubt appear quite dispensable to us, who indulge in no such antics. To the people who practised them, however, they seemed not only of equal importance, but more real, more important, more surely productive of bumper crops than the whole business of breaking up the soil and scattering the seed—inventions which after all must have considerably followed the older rites by which the peoples of the earth first of all insured themselves a luxuriant growth of uncultivated plants.

To supplement science with folk-lore is in the best standing. But properly understood—when the invocation is genuine—it is the folk-lore, the invocation, the ancient ceremonial, which is doubtfully supplemented by technologies. The appeal is from established folk-verities to the doubtful ingenuities of the master minds.

SCIENCE: THE FALSE MESSIAH

Regard, then, the contrast of the smirking pieties of those modern scientists who derive from the "verities" of science their equivocal conceptions of God. Thanks to science, the ancient ceremonials have faded out of the picture. The Lord of Battle is still invoked in the dark days of war; but not by generals. That is a civilian occupation, good for the morale of the citizenry, and excellent propaganda for the war among the "church crowd." But no one takes it seriously. It does not deter generals and politicians from claiming all the credit for having won the war. And do these scientists pray in their laboratories? We doubt it.

When, therefore, we are told that "science is not complete," that "there is still room for faith," we feel that the suggestion is disingenuous to say the least. Incomplete science no doubt is, but not in such a fashion as to lead any one to go behind it to an antique and alien ceremonial. On the contrary, the major premise of science is its undisputed jurisdiction over all our actions. The basis of all scientific research is still the law of universality: conservation of energy and indestructibility of matter. The foundation of all mechanics is the axiom that every bullet finds its mark, and the corollary that bullet holes are made only by bullets. In this technology, what lever, pray, does diety manipulate? Obviously none whatever; nor do the new "discoveries" require any re-assignment of the levers. It makes no difference to the law of indestructibility that in the new physics matter is composed only of immaterial electron-points of energy. Whatever the stuff be called, it does not pass in and out of existence capriciously, nor for moral purposes, but only according to the impersonal principles which "govern" its behavior. Later discoveries disclose new principles, behavior upon a minuter or a still more grandiose scale. But they do not affect the impersonality of the scientific cosmos. How strange it is that the more precise science becomes and the more complicated its mechanics, the more it should require divinity to complete the picture! It is so strange that it is not true. No scientist says it is. He only hints!

Those anxious hints—a pretty comment they are upon our religion! The sublimest conception science can honestly offer to religion is the admission of its own fallibility. No scientific research is final. Every one ends with unknowns. These unknowns can be given names; but that does not make them any less problematical or any more inspirational. Taken all together, they can be called the unknown, or even The Unknown, if capitals add any meaning to a wholly negative expression. What we have to say in that field is, scientifically, guesswork. If some scientists call it guesswork, while others call it God, that must mean that the latter are able to detect a resemblance. Upon the authority of science, then, Unknowability is all that remains of the faith of our forefathers.

PART THREE
SCIENCE TAKES COMMAND

PART THREE

SCIENCE TAKES COMMAND

No TRULY sophisticated man—and a university faculty is not the place to look for sophisticated men!—believes for a moment in the possibility of any general and sustained improvement in human civilization. There is a very good reason for this, which can be stated in terms of logic, though logic is not the most tactful way to state anything. Improvement is an expression of comparison. You can no more talk about improvement except in relation to some particular point of reference than you can talk about distance without having any particular point, from or to which, in mind. Consequently the expression "general and sustained improvement" is logically meaningless. Each separate word can be defined and understood, as a mere matter of language. But when these words are set down together they suck out all the juice of each other's meanings, like the famous weasel words of politics—this, quite apart from the verdict of history which of course finds us innocent on every count. He must be a hardened reformer indeed, utterly sunk in idealism and callous to every prompting of reality, who in these post bellum years can have anything but a grim smile for the notion of perfectibility.

The fact is, academic people are reformers. There is no economist who does not dream of stabilizing currency and arresting the tidal flow of business cycles, nor sociologist who does not contemplate the ultimate elimination of every social "case." Hence the theory of progress, or, as pedants call it nowadays, social control. If each scientist will only reduce his section of the universe to order, it follows that the millennium will automatically dawn. No general conception is required of what civilization will be like upon that happy day. That will be our surprise. Meanwhile, the job has been simply parceled out. You in your small corner and I in mine!

To the sophisticated, it is cause for peculiar wonder that the scientists, who of all people should know best the nature and limitations of their science, are nevertheless more fatuous about it than their own school boys. Science is a scheme of measurement and observation. The elementary textbooks emphasize nothing so much as its impersonality. It has nothing to do with interests and aspirations. It is a machine for taking readings and making calculations with them. What is there in this to encourage the expectation of millennium? Nothing, surely.

But this is only the logical aspect of the case. There is also the dramatic. Tremendous changes have come to pass in the last few centuries. We have solved many problems; and each solution has left us with a new technique, so that we seem to be better equipped for solving others with each century and decade. Not only has the chariot of progress come thundering down the ages; equipped with the sleeve-valve engine and the counterbalanced crank-shaft, it is now vibrationless at the highest speeds, as the advertisements say.

Scientists have no difficulty in believing that science is the driver. But this point, at least, is open to serious argument. Chariots have been known to carry flies as well as drivers. Doubtless science has arrived simultaneously with the rest of the cargo, but in what capacity? Perhaps the mechanical technology could hardly have appeared alone; but then, no more could science. Meanwhile nothing appears less probable than that science held the reins. With incredible dexterity tremendous problems have been solved, as we say. But who ordered the solution of those problems? Whose idea was it that perfection lay in that direction? A quaint and interesting notion which appears nevertheless to have been preposterously wrong. To put it baldly, here we are as a civilization all dressed up and nowhere to go. We imagine that we must be going somewhere because people in that condition usually do—and so we shall. This holiday attire of machinery is going to take us somewhere and no mistake. The point at which we are liable to mistake is in the itinerary. That is a matter on which every one is privileged to guess, and any one's

SCIENCE: THE FALSE MESSIAH

guess is as good as the next man's, or very nearly. Meanwhile, if science is good for anything at all, we may be pretty sure that none of us is going to be able to determine that direction. In its capacity of observer and recorder, science notes that the movement of civilization is determined by the folkways; and the folkways are not amenable to reasoning—not even by scientists. Folkways are not passed by any impressionable assembly to which, perhaps, the enthusiasts for reform might address their exhortations. Folkways—or let us say civilization is an impersonal process like the movements of the stars. We may learn to know its curves or we may not. We may be able to record its movement in advance. In a general way, at least, we can. Of civilization we can say just as we say it of the stars that its future is going to be continuous with its past and in general terms precisely like it. Whatever the winds may be that have deposited this drift, those same winds will continue to blow hereafter; and that drift will continue to form the contours of our social life. Pedants will speculate and enthusiasts exhort. Let them. It is a law of their nature; and the winds of doctrine are a stipulated part of our cosmic weather. The fly buzzes and the chariot rolls on. This is what science teaches about the control of society by science.

CHAPTER XVI

THE LAWS OF SCIENCE

NOT even the most optimistic scientist maintains that science is now regulating human society or even is competent as yet to regulate it. The basis of all our hope is that science is moving in that direction and will probably be able to do so in the future. Like all delusions, this idea is easily accounted for. The first of the modern sciences to emerge were mathematics, physics, chemistry, astronomy, and their associated branches. The discoveries which were made in these fields were very potent. They took the form of mathematically stated "laws." In the history of science, the development has been gradual but steady from these inorganic sciences to such fields as anatomy and physiology and psychology and finally sociology. The social sciences are the last to appear—if they have yet appeared, a fact which students of the inorganic departments often call in question. History, politics, and economics are not absolute novelties quite unheralded by earlier literature; but the conversion of these studies into sciences is strictly modern and obviously incomplete. Thus it appears that the movement of science has been from mass and acceleration

SCIENCE: THE FALSE MESSIAH

to population and finance. Hence our eager hopes of controlling business cycles as we control combustion, and solving the racial problem by a new binomial theorem!

But as pleasant dreams are apt to be, this is a very superficial view of the development of science. It neglects to inquire what science does with its materials and what the causes are of its progress from atoms to societies. Whereas the law of the Lord is perfect, converting the soul, the laws of science are perfect, converting the elements; and the most important and interesting fact about this process of conversion by which scientific analysis proceeds is that it is always a conversion downward. Science moves from the complex to the simple. It reduces everything it touches—substances to molecules, molecules to atoms, atoms to electrons. It began with numbers and planets and inorganic masses because they were already the simplest materials available. Their determining forces and components are comparatively accessible, and the process of analysis dealt with them easily and very surely. Later it was possible to analyze the more complex materials of plants and animals; but the analysis was always in the same direction—downward. Whatever science touches, it eliminates, putting something simpler in its place. No doubt we can still say that the laws of science are perfect, converting the soul; but only in the deepest irony.

The object of scientific thought is to create generalizations, the wider the better. But they are what logicians call universals, though perhaps it would be more sensible to call them qualified universals. Thus the propositions (a) all men are mortal; (b) Socrates is a man; (c) therefore Socrates is mortal, are universals. But the minor promise (b), which should more strictly read "All members of the species Socrates are mortal," is somewhat qualified by the fact that this is a species with only one member; while the sweeping certitude of the major premise (a) loses something of its grandeur through the suspicion that it is only a balancing of synonyms. "Mortal" can be defined only as "having that span of life commonly attributed to men (and other animal species)"; so that "All men are mortal" means nothing more nor less than "All men have that span of life commonly attributed to men"—not a very illuminating statement after all. The laws and hypotheses of science are many of them of extraordinary breadth. They cover all forms of matter and all species of living things. But they are necessarily subject to just such reservations as attach to any universal. Suppose that they are true: they are still quite likely to be broad statements about tiny things, ordinary definitions—the names of things—split in two with "are always" thrust between them. A marble is a spherical pellet of clay or glass. Then marbles are always spherical—a universal scientific law.

Indeed, how else can universals be created? To make a statement affecting "all" things of a certain kind, one must either have seen them all, or one must propose to sort them out on the basis of this peculiar characteristic. For example, before remarking that all the clover leaves in a given patch are "three-leaved" it would be necessary to examine all. But an inspection of a handful is a sufficient basis for the remark that all normal clover leaves are three-leaved. This will be true. But it will be true only because "normal" is defined as meaning three-leaved. No doubt the distinction is arbitrary, and the class established by the statement artificial. The question is whether any distinction is ever fundamentally different, and any classification much less arbitrary.

In its early stages, science was openly concerned with making such distinctions. Scientific research, at the time of Aristotle, for example, or Pliny, was largely a matter of sorting obvious things into convenient classes and sub-classes—species and genera. But we like to think that modern science has entered another phase, and is analytical where ancient science was merely descriptive. The contrast is especially conspicuous in botany where, until comparatively recent times, the whole interest of scientists was in sorting and docketing, in "identifying" new species and placing them in proper relations with the familiar ones. Not only that: the classifications were made on the basis of the number of petals in the flowers, the shape and arrangement of the leaves, manner of seeding, sprouting, and so on. But all these marks of species were defined on the basis of gross characteristics of the various plants, such as any one might observe with the naked eye. When microscopic examination was made possible and the microscopic characteristics of tissues and germ cells were observed, these things at once became the basis of general reclassification. Accordingly, interest in classification has waned, and its place has been taken by interest in microscopic observation. This is analysis. It marks the beginning of "purely scientific" methods.

No doubt this is all very well, and the change is for the good. But it is nevertheless a fact that the present methods of microscopic examination are not infinitely perfect. They might conceivably be improved, and if they should be, further change would probably occur in botany. It would then appear that the late "microscopic" methods of observation were "crude," that the traits observed by their use were superficial, and that the arrangement of the species and their parts worked out under the old dispensation was naïve and requires modification all along the line. In short, all that we say of pre-microscopic botany would then be said of ours. From that point of view, therefore, it would appear that our science has been a mere, arbitrary recording of gross similarities and differences, and a classification of the "facts" upon the basis of our woefully incomplete observations.

Possibly this is a very special case. Yet wherever one turns in the literature of science, the progress of history seems much the same. We pass

SCIENCE: THE FALSE MESSIAH

from one counting and weighing of the stars to another, re-arranging as we go. We descend from molecules to atoms, and from atoms to electrons, leaving successive stages of data and arrangement, pursuing the mirage of analysis from one descriptive level to another. To be sure, we are not merely ranging wide, discovering further genera co-ordinate with those we already know. Each successive penetration opens a new order of magnitude. Ordinarily this is the focus of scientific pride. So far as ingenuity is concerned, this pride is certainly justified. But apart from the triumph of ingenuity, apart from microscopes and other mechanical apparatus which are a by-product of such ingenuities,—considering the new discoveries solely upon their merits as discoveries, it is very doubtful indeed whether the steady regression from the near and obvious to the remote and minute and progressively intangible is ground for rejoicing or for sorrow. A very good case can be made for lamentation. We inquire of nature what makes the grasses green, and receive an answer that is all about chlorophyl, catalysis, and osmotic pressure. We ask what such things are, and receive a reply that is all about molecules and kinetic energy. We inquire about these, and are told of atoms, and then of electrons, protons, and elementary electric charges. No doubt it all hangs together. No doubt each level of investigation "explains" the other. But somehow the answer seems to have moved out from under the original question, so that each successive definition

THE LAWS OF SCIENCE

is after all just a complicated way of saying, "Why are the grasses green? Because they are." Not only do these answers fail to answer: they answer less and less the more minute and precise they get; so that the suggestion inevitably presents itself to the disgruntled questioner that all these formulas involving molecules and atoms are so very precise and invariable because they are very non-committal. The extraordinary precision of science upon such levels of observation is possible because nothing can be seen on those levels but weight and movement.

This suggests more forcibly than ever that in science as everywhere else, things can be asserted positively only when they are devoid of meaning. Atomic relations can be stated with as much certainty as the remark about the mortality of Socrates because—these weights and movements being all we know about atoms, they define what we mean by an atom, and we can therefore say that "all atoms" behave thus and so, with absolute precision. Thus it does not mean the same thing at all to remark that a man weighing one hundred ninety pounds has run a mile in four minutes and ten seconds, and to say that an atom weighing one unit equivalent to the weight of a hydrogen atom travels at such and such speed. We know other things about men which make the first statement significant. But we know nothing about atoms except that they are entities of which we know this. In short, science can speak with great certainty of matters which it selects for

SCIENCE: THE FALSE MESSIAH

notice, thereby astounding us all, notwithstanding our regret that other matters in which we take more interest can not be treated with similar precision. It is, of course, pleasant to know of a verity that Columbus discovered America in 1492, to be so sure about something in a universe of flux; so pleasant indeed as to make it convenient to push into the background the question whether Columbus was the prime mover whose credit deserves the honor, whether the discoverer did hit upon the western hemisphere for the first time or not, whether he touched upon the mainland, an island, or—as Columbus thought—the East Indies, and whether the event is properly assigned to the year 1492. In general it may be said that science reveals with ever-increasing surety a range of facts which take on more and more the character of sheer curiosities.

Of course, atoms and protons are not on this account any less real or the laws of physics any less true. Whenever questions are raised what, if anything, the formulations of science mean, some ardent scientist is sure to leap to his feet with vehement protestations of belief in the "reality" of these particles. "Atoms are real! No one may ever see one. They may never do any one any good. But they are real nevertheless." No one can gainsay it. But suppose that they are earnest as well as real, and that the grave is not their goal—propositions also which no one can deny—what does it mean to say so? Something is real, but what? Atoms are real in the same sense that monocoty-

THE LAWS OF SCIENCE

ledonous plants are real. Without becoming too metaphysical we can nevertheless distinguish between the existence of plant life all over the landscape in every variety of shape and structure, and the intellectual exercise of sorting these plants into arbitrary varieties according as they sprout from cotyledons or not, and if so whether from one or more. Physical objects, also, lie about all over the landscape in still greater variety of shape and structure. These, too, we have sorted; and on the basis of a certain variety of sub-division—a certain class of internal structures—we can call them all atomic, and classify them all, from the monocotyledonous atoms of hydrogen to the atoms, polycotyledonous to the two hundred thirty-eighth power, of uranium. What is real? Doubtless all; the plants are real, and the masses are real, and the botanical divisions are real, and the physical divisions are real, too. But the physical divisions—the atoms— are not real in the same sense in which plants are real; they are real as other orders of classification are real, cotyledonous and otherwise. In this sense, also, the square root of minus two is real: it is a real intellectual abstraction. And the Absolute!

In other words, it is no great pumpkins to be real. Everything imaginable is real in some sense, and true in some sense. But the one great intellectual deficiency of science is that its sweeping generalizations, whetted to a razor-edge of precision, nevertheless unfortunately concern realities which though they may be spread over the length

[217]

SCIENCE: THE FALSE MESSIAH

and breadth of the universe are of an order of magnitude which excludes them forever from the joys and sorrows of human intercourse. Once more be it stated that we are concerned here not with inventions and appliances but with scientific "truth." More simply, scientific truths are not the truths that make men free. They are too true, too universal, too empty of humanity for that.

In the early days of modern science—the boom period, so to speak—a contrary view was held. Thus that famous "law" of conservation of energy was framed very largely to be helpful. It was supposed that such problems as the supernatural would be solved by a simple application of the law. But as time has gone on the situation has become more complex, and the law—steadfast in its universal truth—has receded from the human scene. For example, exceptions to the law have been developed. Somewhere between the radiant stars and the illuminated earths a tremendous head of energy disappears. Other radiations exude from our galaxy never to return. Some authorities have exhibited such cases as instances of imperfect law enforcement. But others have pointed out that they only indicate hitherto unknown routes of the transfer of energy. If the inter-stellar spaces are the scene of a "loss" of energy, some physical process is going on there, under the impetus of this energy, which we simply do not know about. If energy can dribble out of our galaxy, it can also dribble in, to which the visibility of other galaxies attests. In-

THE LAWS OF SCIENCE

deed, the law of conservation can admit of no exception, since it is derived from the nature of energy. It affirms only that energy exists. It does not limit the forms in which it may appear, or from our vision disappear. An exception to the law is excluded by definition. Wherefore, of course, supernatural events need only to be defined as unknown forms of energy to become perfectly law-abiding, and the more literate spiritualists have proceeded to define them so, thereby robbing the famous and inexorable law of all its teeth. Thus it reigns, like the Absolute and the King of England, supreme and powerless, a perfect symbol of scientific enlightenment. For many years it has been known that two plus two makes four. Yet civilization has not been regenerated. What are we to expect of the proposition that one proton plus two electrons makes hydrogen? Doubtless more of the same lamentable unregeneracy.

CHAPTER XVII

SEEING THE INVISIBLE HAND

IF THE laws of nature are the laws of God, it obviously follows that all we have to do to live in obedience to divine command is to discover what those laws are. This, many people think, is precisely what science is going to do for us. In an early edition of the works of David Hume, an obscure editor once inserted an interesting footnote. Hume had been commenting upon the sexual instinct and its marvelous efficacy for continuing the species of creatures who little dream of the great responsibility that rests on every individual. By this simple and unsuspected mechanism, said Hume, God's will is intricately done; which amounts to saying, noted the editor, that God intended a very important matter affecting the continuance on earth of all His children to be a secret, but Mr. Hume found Him out! This remark nicely characterizes the sublime conception of God which science furnishes. It has been supposed in the past that God works in His mysterious ways His wonders to perform. But latterly science has found Him out.

Or so many early scientists have believed. The first contribution of modern science to theology was

SEEING THE INVISIBLE HAND

the mystery of God's ways. No direct traces of divine manipulation were ever found. Consequently the inference was plain: God dwells invisible within the eternal order of things. Since no special action or re-action appertains to Him, the whole order of nature is none other than the divine plan. This was the view of the scientists who laid the foundations of modern astronomy, physics, and chemistry. And from this it is only an easy and obvious step to the idea that the order of nature is holy and must under no circumstances be tampered with.

This was the view of many of the most influential thinkers of the eighteenth century. It so happened that great social changes were taking place in the eighteenth century. At this time, the process of industrial growth, which is sometimes called the industrial revolution, had proceeded far enough to disturb the balance of society. Commerce and manufacture were becoming more important than they had ever been, and the middle, or commercial, classes were likewise greatly increased. Royalty and nobility were at a corresponding discount: hence the social ferment which was troubling the political surface of Europe. In France the "natural order of things" from which all "social contracts" were presumed to date was very much to the fore, and the "noble savage," or natural man unencumbered with the artificialities of sophisticated life, was enjoying great popularity. In England the ferment worked somewhat differently. For one

SCIENCE: THE FALSE MESSIAH

thing, it had already gone much farther there. England was already a middle-class nation with a middle-class government. Commerce and manufacture were not yet in the best odor in polite society; but they were too well established to be in the least apologetic. They were already the back-bone of the nation. Nevertheless, even in England the backbone was carrying an excessive weight of disused members, the vestigial organs of the ancient feudalism. The land-owning nobility were in a position to exact tribute from the growing commercial enterprise of the middle-classes, and did so quite unblushingly. Very naturally, therefore, the order of nature seemed to most sensible middle-class Englishmen to be the middle-class England, the commercial England of their day and age, and not at all that ancient feudal England of which the House of Lords preserved a battered remnant. Add to natural class interest the fact that special privileges could be obtained only by special grants and other interferences with the "prevailing order," and the case for *laissez faire* is complete. The natural order of things is the divine plan; middle-class, commercial England is the natural order of things; therefore middle-class England is the divine plan. Whatever interferes with the divine plan must be stopped; all special privileges do so interfere; therefore special privileges must be revoked and business must be let alone. Thus reasoned the social philosophers of the eighteenth century.

Thus reasoned, for example, Adam Smith.

SEEING THE INVISIBLE HAND

Adam Smith was a very pious man, according to his lights and the customs of the day, though no more so doubtless than most of his university colleagues. He had no notion of abandoning either European civilization or the God of European folk-lore. His attitude toward the land-owning classes was most polite; and he was most reverent in all his references to God. But as a scientist he felt that God was to be seen not in cataclysmic revelations but in the sober scientific study of the order of nature, and that for civilization this meant chiefly the order of commerce and industry. Regarding that order with the eye of a trained observer, Adam Smith believed that he could almost see the hand of God moving behind the inscrutable laws of supply and demand. When business is let alone, he said, and commerce has free play, men seem to be "guided as by an invisible hand" to those very combinations which are best for the public good. Since business is ruled by supply and demand, and since supply and demand are laws of nature and therefore the will of God, business men are ruled directly by the will of God, like the gravitationally-guided stars.

Now as a matter of social policy the French philosophers may have been right in attacking the Bourbon autocracy, and the English philosophers may have been right in opposing governmental interference with the "infant industries" of the new commercial age. But the arguments they employed were nevertheless extremely devious. If any real question had been raised as to what was actually

SCIENCE: THE FALSE MESSIAH

the natural order of things in European civilization, only one answer would have been given candidly. The Bourbon monarchy and the ancient nobility which supported it, the time-hallowed social hierarchy of lords and serfs, were certainly far more truly characteristic of Europe as Europe had always been than was the pin-making industry which so moved Adam Smith to wonder at the efficiency of its division of labor. Here were two forces: political government as old and venerable as civilization itself, and a group of infant industries. The social philosophers of the eighteenth century looked that venerable though declining system in the face and declared without faltering or blushing that the order of nature was the infant industries, that the best government was the one that governed least, and that God though invisible was on their side. Considering that the hereditary autocrats claimed to derive their powers by anointment from a very present deity and that such had been the case from time immemorial, this was tantamount to a declaration of independence from the whole order of European civilization, folkways, folk-lore, and all. Governments are to be seen and not heard; God is to be interpreted but not seen.

This is a strange science and a stranger theology; but its modern counterpart is even more curious. Adam Smith stood at the threshold of modern industry. His *Wealth of Nations* appeared while those very discoveries which revolutionized the textile industry were going on. He had seen commerce and handicraft, but he had not seen machinery, in the modern sense; and neither he nor any of his contemporaries in the faith of *laissez faire* could have had any notion what the mechanical developments of the next century and a half were to be. It is even doubtful if we who can look back upon this period realize its meaning any too fully. But one thing we can hardly avoid seeing, and that is the magnitude, the scale, of the machinery by which manufacture and transportation are now conducted. Doubtless this machinery itself is not sinister. There is grandeur, awful grandeur, in its tremendous size and terrible efficiency; yet, for all Samuel Butler's mid-Victorian vision of the world ruled by machines, it is of course quite impotent without the human touch which sets it going. The machine does not dominate its engineers and firemen.

Nevertheless in a peculiar fashion it does dominate society. It dominates its owners and directors. When transportation is conducted by stage-coach there is no reason to fear our ever being at the mercy of stage-coach owners. As Adam Smith pointed out, any one can own and operate a stage-coach, unless a "corrupt" government prevents; and if the prevailing owners become exorbitant, any one can travel in his own carryall. He can even outfit a coach of his own and go into the business to reap some advantage of the temporary prices commanded in that business. The roads are public, and barring special privilege they are open to all on the same terms. But none of this is true of railroads. Railway lines are not public; they are privately constructed and privately owned thoroughfares. If any individual or even a community of hundreds of thousands of souls feels dissatisfied with the transportation service, there is no such option as mounting new coaches on the existing railway and operating them on more advantageous terms. To duplicate the line of rails takes a very long time, is frightfully expensive, and has become generally quite out of the question. Meanwhile the very magnitude and importance of the equipment by which modern transportation is carried on have made the concentration of control imperative. The various coaches of a stage-coach line are all operated independently. Coaches may be added or removed without affecting the movements of any of the others. This is not true of railway trains. The same organization which controls the movements of one train on a road must control all the others. This is an unavoidable condition of railway traffic. It is similarly true of every industry into which machinery has been introduced. Wherever enormous machines have been installed, a corresponding concentration of control has been made necessary.

The fruit of this control is power—such power as Adam Smith never dreamed of. Much has been written of the concentration of ownership; and this is important. But it is not so important as many of us thought at first, because concentration of ownership is a fact whereas concentration of control is a condition. The ownership of a railway may be concentrated or diverse. This need not mean that the owners—by legal title—either are or are not reaping a golden harvest at the public expense. The railroad may be charging exorbitant rates without the owners getting any of the proceeds; or it may be cutting ruinously close to costs and making tremendous fortunes for certain individuals. It is the men who are in a position to manipulate the road who control, not present ownership, but all future advantage. They can conceal and divert profits from shareholders simply by purchasing their coal and locomotives in such a way as to drain off the funds of the railroad into some other pool of their own. They can ruin one railroad or one section of the country for the purpose of buying cheaply for future re-inflation. This is the sinister power conferred—and unavoidably conferred—by great machines. Its control has become the outstanding political problem of the day.

This, of course, is a totally different condition from any which Adam Smith or his contemporaries, the social philosophers of the eighteenth century, contemplated. Yet the voice of eighteenth century philosophy is still loud in the land—for two strong reasons. One is that it is a very profitable logic for those who are in control of the new industrial system. Whether political government, which is by definition outside the world of commerce and industry being descended from an older order of masters and servants, can ever exert the power necessary to control modern industry is a grave question. There are those who think not. They hold that industry is now too big and too technical to be controlled from outside. They point out that most of the real powers of government are already exercised by the higher orders of business men through the channels of business organization, that in point of power such business men are far more likely to control political machinery than to be controlled by it, that in any case political machinery is impotent to prevail against them. But whether this is true or not, the philosophy of "least government" delivers the world to business, bound, gagged, and hog-tied. Upon the theory that business must be let alone, a theory which business men can proclaim as well as others, the world is their oyster. It is not surprising, therefore, that the official doctrine of modern business is almost wholly derived from the theory of *laissez faire*.

In keeping this philosophy alive, moreover, the business men have the unasked and unpaid help of a considerable part of the "learned professions," including the profession of learning. Bookish men are apt to be antiquarians. They are apt to be more devoted to the past than other men, and more versed in it. They are apt, also, to pay a great respect to "reason," and now-a-days to science. In other words, they are very open indeed to persuasion by the eighteenth century philosophers. The same considerations which appealed to Adam Smith still appeal to them. They also incline to the invisible hand, perceptible only in the light of science, by which God moves in His mysterious ways. They also are susceptible to the esthetic charm of an ordered universe, divinely appointed to ineffable perfection. They also have witnessed the growth of commerce and industry in obedience, not to kings and princes certainly, but to inward laws as potent and as unconscious as the sexual instinct. Therefore they too proclaim the "natural order" and the law of let-alone. Thus the daring insights of men who were fired by a vision of natural perfection have become the amiable pedantries of men who are caught by the charm of ancient books; and the quick sympathy with which men once defended the independent labors of freemen from the impositions of an ancient tyranny has turned into the smug self-satisfaction of a far more potent tyranny.

Whether science ever reveals the divine perfections of God's hidden hand is still open to debate. The whole theory of natural order may have been a natural invention by which men endeavored to draw from science what they were losing from theology. In the field of civilization, the discovery of any such divinely appointed scheme is complicated by two factors not encountered in the observation of the heavens. Civilization changes more than the heavens do, and men have a stake in its concerns as they do not in the relations of the planets. In recent times, moreover, civilization has changed farther and more rapidly, perhaps, than

SCIENCE: THE FALSE MESSIAH

ever before; and the stakes are now correspondingly greater. To proclaim in the face of these facts that business is natural and control is artificial is not science. It is not theology. It is no part of our ancient folk-lore. It is just plain human prejudice and self-interest. As such it may be a prejudice to be reckoned with. The folk-lore of the future may be germinating in it. In the modern industrial corporation we may be viewing future society in bud—or we may not. In any case, that is another story.

CHAPTER XVIII

SOCIETY IN THE LIGHT OF REASON

To REGULATE society by the laws of science is a laudable ambition—theoretically. The idea is a captivating one. Like the theory of the order of nature which needs only to be let alone to realize the kingdom of heaven upon earth, it has a certain esthetic charm. It would be nice if things were like that. We can stamp out a disease by discovering its cause, as for instance mosquito bites, and then eliminating mosquitos. Why not stamp out crime by the same process? Crime, let us say, is a product of poverty. We have, then, only to stamp out poverty to prevent all crime. There is only one flaw in this reasoning, and that is that poverty is not like mosquitos, something which can be extirpated, leaving all the rest of the visible world very much as it was before, but minus mosquitos and minus yellow fever. Poverty is a very extensive and ancient institution. It is the nether boundary of affluence. Stamp out poverty and you stamp out affluence as well. Cut away both, and you have amputated civilization. In a very real sense, poverty is one of the elements which constitute civilization. So is crime. Crime is the nether boundary of

SCIENCE: THE FALSE MESSIAH

moral and civic rectitude. Eliminate it and you eliminate civic virtue. There is no such thing as decency when there is no such thing as vice. In fine, the disagreeable aspects of civilization are just as much a part of the order of nature and just as definitely ordered by whatever laws there are as any other aspect.

Scientifically speaking, this is also true of mosquitos—which only means that we do not exterminate mosquitos as the result of scientific insights. We were resolved upon it for reasons of our own as an expression of our prejudice against dying of yellow fever before we knew that mosquitos have any connection with that malady. Science applies only the means of achievement. The problems of civilization, on the other hand, are questions of what we want to achieve. Science has devised no technique for solving them. Whatever passed for solutions in former times was a product of the folkways, that is, of customs and habits that had always in some fashion got established. Reasoning from the past, the inference is that we can break away from one set of habits and the "superstitions" associated with them only by falling into another set of habits. What place is there for science in this cycle? In such a medium science can be effective only by abandoning its character of technician-in-chief to the mechanical arts and assuming the pontifical robes of folk-lore. In that character it can pour the anointing oil upon some "newly" established scheme of folkways. It can reveal our "infant industries" as the order of nature, and itself as a new and sublimer conception of God as the invisible hand. All this is quite effective—as a means of lending countenance to commerce. But it is not science in the "modern" sense; it is folk-lore in a very ancient sense indeed. On the other hand—and this is the real tragedy of science—to be itself it must put off its robes of omniscience and repair to the machine shop. There it can be modern, and efficient, and mechanical. But there, alas, it can do or say nothing that is in the remotest degree applicable to the problems of civilization.

We are just beginning to find this out. A generation or two ago, we began to realize that our infant industries were becoming lusty enough to tyrannize over the whole family. Simultaneously it occurred to us that human sympathies must have played a large part in the "intellectual" process by which the eighteenth century philosophers hit on commerce and manufacture as those departments of life most beloved of God the Invisible King. We could not accuse those grand old scholars of that kind of personal interest which was only too evident in the special pleading of public-be-damned business men. So we called theirs "unconscious bias" and when we had dissected it out, we labeled it "preconception." With these things in mind we determined that industry may require regulation after all; and we proceeded to regulate it with minds well cleared of the *laissez faire* preconceptions of an older and less scientific day. Our procedure should

SCIENCE: THE FALSE MESSIAH

be based on facts; and we went on to produce them in what is now generally known as the "era of muckraking." The stench was very great. We were therefore led to assume that the method was effective. We were at long last truly scientific. We were not formulating theories; we were exposing facts.

To our dismay and chagrin, however, all has not worked out precisely as our scientific calculations led us to expect. Gradually it began to appear that facts, even the most factual facts, are not self-operating. They require to be cranked up by some human driver before they will go anywhere; and this driver may be so very human as to influence the direction they are given. We could no doubt control our industries. But what is control? No one could possibly advocate a blind control which would exterminate all commercial combinations evenly and impartially, leaving nothing in the land but isolated business men. What we required was judicious control in the best interests of society. But the best interest of society, like prejudice against yellow fever, is a matter of tradition, and traditions are always subject to interpretation. In short, we found that the most eloquent facts mean different things to different men, and that what happened as a result of our determination to control our now stalwart industries was very largely affected by the human predilections of nine elderly men upon the Supreme Court of the United States. Preconceptions might be supplanted by facts in our array of evidence; but they could hardly be excluded from the breasts of even the most august court. Some members of the court might be quite up-to-the-minute public-be-damned men of the world in their attitude toward these matters. Others might be sincere antiquarians, gentle and bookish octogenarians of pre-Adamite philosophy. Others might be warm-hearted sympathizers with the under-dog of the modern proletariat. But none could be scientific. That would be impossible. Moralists say that to understand everything is to forgive everything. Perhaps. But in science the rule is to know all the facts and to have no opinion. Opinion is a substitute for facts. It is the point at which facts leave off. Now facts or not, opinion or not, a court has got to get something done. It has got to exercise control. It has got to exceed facts by proceeding to acts. Those acts will be human acts in spite of everything facts can do. They will be based on the facts; but they will be the effect of those inanimate facts upon human preconceptions.

Thus we have discovered the human equation in social control. We have learned to our sorrow that facts propose but courts dispose. In the meantime, however, we have been more impressed than ever by the methods and techniques of science. The period of greatest social disillusionment—the period of great reforms gone awry and ending in international and industrial strife upon an unprecedented scale—has also been the period of greatest scientific achievement. The oldest and

SCIENCE: THE FALSE MESSIAH

best established sciences, which we had almost begun to address in the past tense, have suddenly revolutionized their fundamental laws; while the newer ones, such as the various branches of biology, have made equally rapid strides toward maturity. More than ever before we feel now that science is mighty and will prevail. Consequently we have been more strongly convinced than ever that what is wrong with our social science is an insufficiency of science. In our chagrin over the persistence of human factors in the conduct of civilization, we have turned for compensation to a refurbishing of inanimate facts.

These facts are to be of various kinds; but especially they are to be psychological. The reason for this is that psychology is the science of the human equation. We have high hopes that by approaching these problems of civilization with a real scientific knowledge of the springs of sentiment in the human heart, we can dispose of them scientifically, consulting the heart only in the laboratories of psychology. In this fashion we shall be able to circumvent the preconceptions which so often influence even social scientists in the statement of their laws. We shall analyze preconceptions and provide against them. We shall similarly analyze the motives by which men in society are moved. The soft spot in eighteenth century philosophy was the theory of motives. Adam Smith and the others supposed men to be moved by quite simple considerations of economic self-interest. But later psychology has showed their economic man, their Robinson Crusoe on his desert isle, to be a far more complicated creature than they supposed. Very well; let modern psychology go on and exhibit the true economic man in all his complexity of motive. When it has finished, we shall have a sound scientific basis upon which to solve the problems of civilization. Social science, as we say so often now-a-days, waits upon psychology.

This is a large responsibility for psychology, which has its own preconceptions to deal with. In particular, it has been very much occupied with getting rid of the preconception that psychology is the description of the soul. As any one can see, the soul is a perfect storm-center of preconceptions. To be scientific, psychology also has to be objective, to deal only in established facts, to avoid ancient beliefs. Consequently psychology has become a laboratory science, and has more and more excluded from consideration anything which can not be examined with instruments. The result of this has been, as every one knows, that psychology has become more and more a branch of biology and less and less a department of philosophy. The more it has had to say about "the organism as a whole," the less it has cared to hazard an opinion about man, the mysterious author of civilization.

Surely there is no reason for disparaging the science of the organism. If psychology and neurology between them succeed in unraveling the structure and working of the nervous system, we

SCIENCE: THE FALSE MESSIAH

may be in a position to cure many diseases which now resist all treatment. This will be no small gain. But curing epilepsy and preventing dementia præcox are very different matters from solving the mysteries of human motives, just as different as stopping yellow fever and preventing crime. Psychology may become more scientific by such procedure. It may even become—what William James hardly dared to hope—a regular branch of science. But it does so at the cost of ceasing to peer into the human soul. The more it knows of reflexes and secretions, the less it is able to choose among them. If "all human behavior results from conditioned reflexes," what conditioned reflexes should Supreme Court judges have? This, it will be seen, is not a scientific question.

If we turn the search-light of science from the human organism to man-in-society, the result is just the same. We get an accurate, detailed, and very disillusioning picture of how our vaunted civilization works. But we get no help. The fundamental "law" of social psychology seems to be that all human behavior is formed by pre-existing human behavior. This is the law of continuity. In adopting such a statement social psychology moves over into the neighborhood of anthropology just as psychology has courted physiology. Anthropology is the natural history of civilization. Instead of reflexes and secretions, it deals with folkways and folk-lore. The problems of anthropology are to identify the folkways and folk-lore of any given

SOCIETY IN THE LIGHT OF REASON

people as psychologists identify conditioned reflexes, and to trace them to their historic source. The fundamental law of anthropology is that all culture comes from culture, the law of continuity in another form. Anthropology is a science. The anthropologist may be able, by amassing tremendous quantities of evidence, to prove that a given culture trait, such as jazz music, was derived from such and such a source. He may prove that it comes from negro folk music, and that this came from ancient African ceremonials connected with tribal life in the African jungle. This enables us to make whatever inferences we like. We may then proceed to exterminate jazz by exterminating the negro race, or more humanely by making all colored people deaf-mutes and forbidding the publication and performance of tainted works. But there is no scientific authority for such procedure. The whole meaning of the law of continuity is that to science all culture is alike, just as to science all microbes are alike. Invidious discriminations among them are due to human prejudice. So are invidious discriminations among folkways.

But anthropology, which seems to be the end of the series of sciences to which we can appeal, is rather less amenable than the others to human use. This array of sciences suggests the experience of the old woman whose pig refused to get over the stile. She appealed to a stick to beat the pig, and then to a fire to burn the stick, and then to water to quench the fire, and then to an ox to drink the water,

SCIENCE: THE FALSE MESSIAH

and then to a butcher to kill the ox, and then to a rope to hang the butcher, and then to a rat to gnaw the rope, and finally to a cat to kill the rat. The cat, it will be remembered, made conditions. But they were conditions with which the old lady was able to comply, so that very shortly the cat began to kill the rat, the rat began to gnaw the rope, and so on with the inevitable salubrious result. The appeal to reason in the manner of modern science has led from pillar to post and seems to end with the science which reveals the inmost mysteries of civilization. But unlike the cat, that one is less open to persuasion than all the others. Most sciences maintain an attitude of benevolent neutrality toward human aspirations. If we want to arrive at certain ends, science has nothing to say against it, and may be able to provide a vehicle. We can still hope for some other science to provide the signboards directing us where to go—until we reach anthropology. There we learn all too clearly that whither we go and why are matters toward which science may display a polite curiosity; but they are no more to be extracted from science than heavenly salvation. The whole meaning of the law of continuity is that they have got to come from us, from our pre-existing organized behavior. In short, human life moves by preconceptions and can proceed in no other way. This is what we learn from science.

Whether any advantage comes of finding out such things is an open question. We have a saying that knowledge is power. It is one of our favorite preconceptions in this scientific age. No doubt there is some truth in it. Obviously many different kinds of knowledge lend power to many arms. But here we are dealing with an extraordinary bit of knowledge: the knowledge that in civilization folkways are power. What power is lent by the knowledge that where human aspirations are concerned knowledge is impotent, and to whose arm? Perhaps some scientist can answer this question, and perhaps not.

CHAPTER XIX

ASTROLOGY AND PSEUDO-SCIENCE

FROM time immemorial one object of scientific research has been to discover in nature secrets of tremendous import to mankind. The interest of early and even primitive civilizations in astronomy is of this character. We are so amazed by the ingenuity of early astronomical devices and the incongruous accuracy of so many of the Chaldean or Mayan observations that we overlook their purpose. Such things seem incongruous to us only because we can scarcely understand why peoples who have no need for a universal system of astronomical time and no interest in an astronomical theory of tides should take the trouble to develop this one isolated science. They had no general scientific enlightenment, no Copernican revelation to be up-held against contrary dogmas. Indeed, their astronomy usually existed side by side in apparent amity with sun-worship or some other form of astrology. Precisely: their astronomy was astrology. It was a branch of their religion. The expectation of their scientists was the discovery in the spots of the sun or the relations of the planets of some occult principle for the guidance of human destiny.

ASTROLOGY AND PSEUDO-SCIENCE

We assume that science has abandoned such pretensions. Has it? In the preface of his book, *Body and Mind,* published in 1911, Professor William McDougall, of Oxford and Harvard, defined his interest in his scientific problem as follows. A belief not only in the soul but in its immortality in future life has been a feature of all the successful civilizations of the past. It is my opinion, said Professor McDougall in substance, that this belief is being seriously weakened among the peoples of modern European civilization. This is a very grave danger. It is very questionable indeed if the disappearance of such beliefs, with their tremendous stabilizing power, would not mean the speedy dissolution of a decadent civilization. Now for myself, he went on, such beliefs are not necessary. I am not conscious of any vital dependence upon the future life as a restraint from vice or a spur to righteousness. Probably this is true of the better educated classes generally. But the educated class is in a small minority. The great mass of society is not suitably equipped to face life without these stabilizing faiths—without the knowledge of the future life and what it may contain. Now science is the authority to which these people look to-day. Only what science demonstrates is now universally accepted. Therefore it behooves the men of science to discover and demonstrate as fully as may be the nature of the soul and the conditions of its immortality, and that without a moment's more delay.

No doubt this is a rather unusual statement. It

SCIENCE: THE FALSE MESSIAH

has few parallels in the writings of modern scientists in good repute. To attack it on the ground that it is quite exceptional, being merely the expression of an individual idiosyncrasy would be very easy; but it would be just as easy to defend it on the ground that while such candid statements are unusual among scientists, the sentiments expressed are not: on the contrary all the reconcilers of science and religion and signers of credos attest the stalwart faith of scientists of eminence who hold just such views as this. What is the difference between saying that science ought to prove the immortality of the soul and saying that science does prove the existence of God and even furnishes a sublime portrait of Him, as the forty signers of the Millikan Credo did? To be sure, most of those men would almost certainly insist that their expression of faith was not a definition of the chief aims of science. They meant to state not that divine revelations are the object of research but that they are a most salubrious incidental effect. Science is mainly interested in the truth about electrons; but incidentally it does turn up some interesting data about God. Certainly this is not Professor McDougall's position; and his defense of his more extreme attitude would no doubt be that he is more candid and consistent than the others. He has avowed his motives openly: for the good of civilization he wants science to discover a mystic principle.

Is this motive really odd or really perverse? History, at least, denies the latter charge. It is only

ASTROLOGY AND PSEUDO-SCIENCE

very recently that scientists have taken any great interest in anything else but mystic principles: philosopher's stones, elixirs of life, occult decimals, miraculous eclipses and the like. And have these interests entirely disappeared? In that case, let us ask ourselves why people are interested in the relation of sun-spots to earthly weather, or in the reception of radio waves from the great open spaces of the cosmos, or in the meaning of the mountains of the moon. Why the perennial anxiety to communicate with Mars? Why the tremendous flurry of excitement over the announcement, in 1919, by the British Royal Astronomical Society, in solemn conclave assembled, that certain stars were out of place as predicted by Albert Einstein in works of such awful obscurity that only twenty in the world could read them? Why is it a matter of such tremendous import that the whole world, yea, all that is therein, is composed of one all-constituting substance—electrons?

We shall be able to find the answers to these questions only by searching our hearts ruthlessly. Scientists will meanwhile tell us that as a matter of fact they have very little professional interest in Mars. Among astronomers Mars has no special importance except as a never-failing subject for popular lectures. As to relativity, every astronomer and physicist knew the main outlines of the theory years before the eclipse of 1918, and knew that the importance of the observation of that eclipse was the opportunity it afforded of checking certain in-

SCIENCE: THE FALSE MESSIAH

teresting predictions. It has been said that every physicist, mathematician, and astronomer in the world has since that time written a popular exposition of relativity to supply the unprecedented demand for works on this scientific subject, and that most of these books have been published. Scientists would defend this course, which might otherwise seem to resemble opportunism, on the ground that the scientist must take advantage of every opportunity to educate the public in so important a matter as science, and he must begin wherever he has a chance. A rift in public apathy at any point is too important to be missed. But most scientists would admit at the same time that what laymen are likely to glean from these popular expositions falls far short of training them as scientists.

We are now approaching the disputed soil of pseudo-science. Between authoritative science and ignorance pure and undefiled lies an intermediate territory where science shelves off into popular science, followed by the pot-holes of pseudo-science and the bogs of quackery. Where one begins and another leaves off is wholly a matter of opinion. One way of bounding these areas more exactly is to say that science is what an accredited member of a learned profession writes solely for the eyes of his fellow members; popular science is what an accredited member writes for the general public; pseudo-science is the same thing written by an outsider; while quackery is the expression of an opinion

ASTROLOGY AND PSEUDO-SCIENCE

on a controversial subject for a general public by an outsider. There is one objection to this system of boundaries, however, and that is the extreme difficulty of distinguishing one work from another. A scientist of real eminence may not only write for a general public, as Thomas Huxley did, but he may even write silly nonsense on highly controversial subjects, as Sir Oliver Lodge has done. A popular lecture by Huxley may be quite indistinguishable from a similar lecture by, let us say, John Burroughs. Is Huxley's work to be assigned to a higher elevation simply because of its author's greater scientific attainments? A popular work by Princess Zozo of the Inner Silence may be not a whit sillier than a popular work by Sir Oliver Lodge. Should Lodge's stuff be treated with the respect due his earlier physical researches?

These questions are important because it is impossible to probe the meaning of science without asking them. How, otherwise, can we know whether the object of our investigation is science, pseudo-science, or mere quackery—which is worth no one's time to criticize? For example, scientists and spokesmen of scientists tell us that science reveals the laws of nature and that we must obey them or pay severe penalties therein severally and with great gusto enumerated. "The laws of health" are a case in point. How often are we not told with the solemnity the occasion merits that science declares unto us the laws of health and that whosoever sets those laws at naught the same shall suffer for it

SCIENCE: THE FALSE MESSIAH

even unto the third and fourth generations—if any. Indeed, this particular commandment has been repeated until usage has sanctioned it. It can be uttered in the best society. This much we accept as science. The law of population growth is not quite so well received. We are accustomed to discussing the subject; every one seems to agree that there is such a law; but precisely what it is can not be stated without challenge. A prediction or a warning based on some one or another of the proposed rates of increase is, perhaps, popular science—not quite accurate and authentic, but still by no means to be sneered at. The law of Nordic superiority, and the commandment to exclude immigration of the brachycephalic and otherwise objectionable "races," are quite another matter. Here, apparently, we are dealing with statements actually in disrepute. Reputable authorities deny that there are any pure races and that any clear evidence of Nordic superiority exists. Much of the discussion of the subject is by persons in questionable standing. The whole matter is clouded with suspicion. This, we may assume, is pseudo-science.

Here again we seem to be distinguishing only between the things that are said and among the people who say them. By accepting the laws of health, we assume that science establishes some laws for human guidance, only boggling at certain other laws which seem less securely proved. But what about the laws of health? Are they parallel to the law of gravitation? Emphatically not. The

ASTROLOGY AND PSEUDO-SCIENCE

presumption is that we can break the laws of health, though if we do we must pay. Does any one expect to break the law of gravitation? In a rhetorical sense we might say that aviators do. Newspapers sometimes speak in this fashion: "Aviator defies gravitation in hair-raising stunt." But this is pseudo-science of the grossest kind. In the scientific sense gravitation is a law only because it defies infraction.

This suggests that the laws of health, familiar as they are, may nevertheless belong on the same bench as the weight-defying aviator. Of course that is the case. If physiology is a science, as some claim, it follows the plan of physics. In physics, a scientific law is not a statement of what falling bodies must do, but of what they actually do—every time. The same is true of physiology. The laws of physiology state not what organs must do, but what they do do. If these matters of general health are laws at all, which may be doubted, to be scientific laws they follow this universal formula: excessive addiction to carbo-hybrates produces obesity; indulgence in strychnine produces death. There is nothing in the law of gravitation to forbid people sitting under apple trees or jumping off the Woolworth Building if they wish. There is likewise nothing in any law of physiology to discountenance or in any way discourage the consumption of strychnine for suicidal purposes. Physiology and its laws are served as inescapably by ascetics fasting in the wilderness as by middle-aged females digging their

SCIENCE: THE FALSE MESSIAH

graves with their teeth. Eat and grow fat; starve and grow thin—thus speaks physiology. As it happens, different fashions reign in different places. The boyish figure is popular in France and America. Buxom contours are preferred, according to tradition, in Turkey and Zululand. Suicide has always been discountenanced by the Christian church; yet in certain circumstances it was not only honorable but obligatory in the codes of Roman patricians and Japanese Samurai. Do the laws of health vary from time to time and from town to town? And in that case must we expect to find localities in which apples never fall out of trees? Behind these questions there lies an awful truth which we can not avoid facing any longer: the laws of health are not scientific laws at all. No law which commands obedience and threatens punishment is a scientific law. Such a thing is the very essence of pseudo-science. We shall never get a clearer definition of pseudo-science than this: whatever represents either science or the forces of nature, as a sheriff's posse waiting to nab us at the crossroads, is quackery and no mistake. Subject matter makes no difference. Authorship makes no difference. If we say that it is a law of mathematics that whosoever says two plus two makes five will be arrested for falsification of accounts, we are naturefakers. Mathematics says nothing of the kind. The laws of mathematics are just as serviceable to swindlers as to everybody else. It is we who are squeamish, not science.

ASTROLOGY AND PSEUDO-SCIENCE

In taking this position, however, we seem to have departed somewhat from our historic picture of science the astrologist, science the handmaiden of theology. But science itself has made the departures—or has at least been struggling to break away. Be as cynical about it as we wish, astronomy is different from astrology; chemistry is not a mere continuation of alchemy. The aims of modern science are not the aims of the ancient nature-delvings. Modern science has become interested in machinery, in mechanical process generally, and in the formulas by which the wheels go round. Its tremendous development has been the result wholly of its specialization in that direction. Now it is the nature of wheels, planets, and electrons, that they turn equally for the just and for the unjust. Consequently this specialization requires a complete break with the past of astrology and alchemy, and a complete break with theology and social welfare past, present, and future. It is impossible for science to build telescopes and electric wave detectors and also at the same time and by the same technique to discover the philosopher's stone, and demonstrate the immortality of the soul, and exclude immigration, and restrain population, and keep fat women from eating too much. After all, science is not a magic wand.

If Professor McDougall thinks it is, his mistake is historically pardonable. Science has never quite achieved that distinction; but it gave up trying only very recently, so recently as to make the confusion

SCIENCE: THE FALSE MESSIAH

of aims and methods almost a certainty. So natural is the confusion, that every one suffers from it a little. We have all fallen victim to the "laws" of health; yet those "laws" are just as preposterous as a law of immortality, only a bit more homely and familiar. If science could guide us to bodily perfection, why not to spiritual immortality? The fact is that it can do neither. It can describe what happens at certain points with uncanny accuracy and minuteness. It can provide instruments for bringing about various effects never dreamed of before in the history of man. But to determine what had better happen or to oblige us to bring about salubrious effects, science is completely impotent. Its very successes in the realm of facts and means have been made at the sacrifice of all claim even to the slightest concern with ends. The guidance of human destiny may have been mismanaged by the Pythagoreans, with their mystic numbers and their music of the spheres; but it was their sincere concern. The only attitude toward human struggles appropriate to modern science is serene indifference, the indifference of the dynamo, the indifference of the mechanical calculator.

CHAPTER XX

SCIENCE WILL PROVIDE

WHETHER we admire the fruit of science or not, we are all so astounded by its size that we are very nearly blinded to the character of the scientists and the actual processes by which they have produced it. Salubrious or not, their works have been titanic. Very naturally, therefore, we think of the scientists as Titans, and credit them either individually or at least collectively with almost unlimited powers. Men have never seen their like before. They have us hypnotized. In this state of temporary suspension of the critical faculty, we find it very easy to believe that science will provide. Our oil may be exhausted: science will find another way; our population may multiply like flies: science will provide food. Science will prevent war after the formula of Alfred Nobel by making it too dangerous to risk. Science will stamp out disease after the formula of Pasteur by antitoxins for bacteria we can not recognize, or after the formula of Ehrlich by filling our veins with arsenic, or after the formula of d'Herelle by cultivating in ourselves the diseases which slay bacteria. Meanwhile science will at last produce gold from lead to make us all rich, and remove the

SCIENCE: THE FALSE MESSIAH

curse of Eve to make us all finally happy. Some of these things may occur. But that is not the point. Many other things will almost surely occur at the same time, most of them quite unexpected and even undesired. The important question is whether science, considered only in its capacity as technician-in-chief to civilization can actually be counted on to deliver the goods when, where, and as desired. Science will provide—but what? What we have ordered? What the scientists in their superior wisdom may think we need? Or what impersonal circumstances accidentally allow?

The picture of the science-superman is too familiar to need describing: the photographic observer, the rigorous thinker, the inspired imaginer of hypotheses, the devoted and under-nourished servant of humanity, and all the rest. If any one needs to renew acquaintance with this paragon, he will find a goodly company of him in a recent book called *Microbe Hunters,* by Paul de Kruif. These portraits possess also this added virtue—they suggest the man behind the mask. Take for example de Kruif's picture of the Italian bacteriologist, Spallanzani. He had just demonstrated a pretty point in the life history of protozoa. Then, says the author, when any other man would have been only too well satisfied with his result, what did this man do? With perverse ingenuity, he set to work devising experiments to prove his own work wrong. Ah, there we have science! There we have real devotion to the truth! There we have the subordination of

SCIENCE WILL PROVIDE

every human instinct to the service of civilization! So says the author.

What he fails to mention is the strange behavior of bookkeepers who, having added a column of figures upward and found the sum, actually—so great is their devotion to the truth!—add the same figures downward in the quaint hope of detecting themselves in error. Of course, to be fair to the rest of the human race, we are bound to note that if the book-keeper made a mistake his employer's accountant would almost certainly run it down, with disastrous results for the bookkeeper. But if we are to make that allowance, we are also bound in justice to the bookkeepers of the world to raise the question whether any one else might ever have checked up the demonstrations of the noble Spallanzani. It is impossible to deny that such is the case. We need not feel any less kindly toward the great scientist on that account; but we are bound to note that he had as great a personal stake in being careful as a man can ever have in anything: if his demonstrations had been inaccurate, and if any of his numerous opponents in the world of science had found it out—as some of them were sure to do—that would have been the end of Spallanzani. In doing as he did, therefore, he behaved exactly as any sane man or child above the age of five would have behaved and does behave every day of his life: he makes sure that the lid is on the jam jar before he says "I didn't do it."

This is what all the peculiar virtues of the scien-

SCIENCE: THE FALSE MESSIAH

tist amount to; they are the ordinary acts of ordinary men magnified ten thousand times by the projecting lens of fame. Not that a very able scientist is the precise equivalent of a very ordinary bookkeeper in the skill and effect with which he does his stuff. An able man is by definition rather abler than an unable man. But to suppose that a scientist is more alert than a prize-fighter, or more observant than a painter, or more rigorous than a grammarian, or more ingenious than a woman in pursuit of a husband, is to think nonsense.

To know that one is thinking nonsense is never pleasant; and to do so without knowing it may still be inconvenient. This is peculiarly the case when we are considering what leads scientists to make the discoveries they do. Thinking of their greatness we are naturally inclined to suppose that they wanted to solve these tremendous problems and so kept at it until they did. Looking at their actual behavior, nevertheless, we almost invariably see something as different from that as can be. Another of de Kruif's heroes supplies a beautiful example. One of the greatest scourges of the human race is a disease called syphilis. Whether the scourge is deserved or not and whether to attempt to cure it is immoral or not are questions that need not detain us here. The fact is that many people are afflicted with a most disagreeable and dangerous malady, and many people therefore desire to be cured, and many others sympathizing with them are ready to hail the conqueror of this disease as a great benefactor of

[256]

the human race. In our enthusiasm, we assume as a matter of course that he wanted to do precisely what he succeeded in doing. We honor him not only for the skill of his achievement but for the superb heroism of his intentions.

But examine the record. It seems that this hero, Doctor Paul Ehrlich, was a biological chemist. He was fascinated by the technique of staining cultures of bacteria for microscopic study. This is a peculiar field. The organic dyes are of almost incredible complexity and instability. Their behavior also is well-nigh miraculous. One dye will stain a certain microbe but not its nearest neighbors; another differing but a whiff from the original will stain the neighbors but not the first-stained microbes. Working in this field Ehrlich became a fanatic on the subject of dyes. He conceived not only the possibility of finding dyes for every microbe, but of using the dyes with their miraculous selective power to poison the microbes in living bodies. For the latter purpose he elected to work on a certain "trypanosome," which he selected because it was large and easy to identify and because it could be inoculated into conveniently small experimental animals. He worked at this problem with these animals and this trypanosome and his precious dyes for a long time, patiently, carefully, learnedly, and no doubt with the most consummate skill. In the end, contrary to the expectations of some of the most eminent scientists in his field, Ehrlich did produce a dye which had the desired effect: injected into the veins of experimen-

SCIENCE: THE FALSE MESSIAH

tal animals it killed the teeming trypanosome without apparently harming the infected animals. This was a great triumph for creative imagination and skilled technique; but syphilis has not yet come into the picture.

According to de Kruif that disease occurred to Ehrlich only after his successes with the trypanosome. In his hour of triumph he bethought him of what this demonstration might lead to, and then he recalled the pronouncement of another bacteriologist to the effect that the spirochete of syphilis closely resembled and might possibly be related to the very trypanosome on which—for quite different reasons—he had just been working. Thus it was suggested to him that he try his potent organic compound of arsenic on patients suffering from syphilis. This was done. The results, as every one knows, were miraculous; and thus, in an instant, was created the famous "606" treatment for this unpleasant malady. In concluding his account, de Kruif dwells with almost rural naïveté on the marvelous simplicity and diffidence of his hero. This epoch-making scientist, this Titan, was actually known to reply to the praises of his admirers that after years of work he had had one single minute of good luck. There is the true genius for you, warbles his biographer. Think of a man who had done that, who could say that! What the author again, as ever, overlooks is that it was true. To have demonstrated this point in dyes was to accomplish what he set out to do, and important enough, science knows; but to have found a cure of syphilis to boot was luck, pure and unadulterated.

How great a part this kind of luck has played in the development of science no one can say. The history of science is, second only to the history of invention, the least known aspect of our past. Mountains of research have been heaped up to prove and disprove the celibacy of the Virgin Queen and Madame Pompadour's influence upon European history, while the factors which led to the introduction of the differential calculus and the dynamo remain almost totally obscure. But this very fact is significant. Persons may or may not have been the potent arbiters of our social and political history; in science they fade into the background, and what we see instead is a string of formulas, one leading as it seems inevitably to another, or a species of machines proceeding from gross and clumsy forms to finer and more subtle contrivances and carrying along with them a progressive string of inescapable observations. It is not possible that we should have the dyes we have without some one's noting that dyes are to a degree poisonous. The discovery might have been made in a less spectacular performance. Most discoveries are. That is why we never hear of them and never consider them discoveries until something spectacular does occur—which we straightway do brand as "epoch-making," disregarding the long period during which the new epoch has been incipient. What we need to know and do not is just what were the elements of inven-

SCIENCE: THE FALSE MESSIAH

tion and formula without which any particular discovery could not have been made and possessing which made that discovery then only a question of time and tide.

When we do not know this of the past, how is it possible to know it of the future? Clearly it is not possible. This is why institutes and foundations and million-dollar funds for research fellowships are not always as productive as they seem to have a right to be. In a certain sense every discovery must be made by some man; and while to humanity his name may be only an accident of time and tide, to him and to the people who are backing him it is a matter of some importance. His backers may know (or think) that a certain demonstration is now in order; and yet the man they have picked may not have the skill to bring it off. Granted the inevitability of scientific progress, the discoveries do have a way of eluding us as individuals. This is the problem which confronts the heads of the foundations: how to know the winning horse! Incidentally, the same problem confronts all such institutions, universities included. The creation of new endowment for research does not bring men into existence to do the work: the various institutions compete for the men they already know. And as the directorates of research foundations are for the most part university presidents, we are confronted with this dilemma: if those men knew what scientists are fated to make discoveries, they would keep them on their own faculties; if they did not know, how could they now invite them to research foundations? Are we to believe that university presidents improve in wisdom when they sit on the boards of these institutions? Or must we suppose that scientists are improved by body-snatching from one institution to another? No, we are forced to the conclusion that the multiplication of scientific salaries and materials means only that in the long run more aggregate man power will be flowing in those channels. This may or may not mean more discoveries per head, or even more per year.

That it should mean the immediate solution of any predetermined problem is quite preposterous. That is not the way problems are solved. Nothing has made this more painfully apparent than the research on the disease known as cancer. Cancer is a most unpleasant, most agonizing, most dangerous disorder wholly without moral or social sense. It falls upon high and low, just and unjust, with superlatively appalling results. Consequently many enormous sums of money have been given to establish research foundations of all kinds for the study and cure of cancer; so much so indeed that it is a saying among biologists that to get a free education or to prosecute a problem you have only to show your keen interest in cancer and all the rest is added unto you. Now it would doubtless be going too far to say that the result of all these efforts has been nil. The authorities of these institutions are most emphatic in their assertions that much has been done. Under certain rare and special conditions the

SCIENCE: THE FALSE MESSIAH

disease can sometimes even be arrested, though not so surely as to convince insurance companies. But no honest man who knows the facts can deny that, measured in terms of present knowledge and treatment of this disease itself, the results achieved are incommensurate, widely incommensurate, ridiculously incommensurate with the aggregate of effort expended. Just what it is that we do not know or can not do, the lack of which sets at naught all our efforts to get at cancer, no one can state with any sort of definiteness—obviously. Consequently we are quite at a loss to know what to do.

Our helplessness in such an event as this is most significant. Since we do not know what to do to cut this Gordian knot, clearly we can not tell who may or may not be doing it perhaps at this very moment. For all we know we are balking his progress by our very zeal. We pile up endowments for cancer research. We draw men off other things to that, and divert their equipment to our use. In this way, perhaps, we hinder and defer if we do not permanently prevent the very work without which all our to-do will be fruitless.

Such is the situation in which we look to the laboratories from whence cometh our help. It is very doubtful if any major discovery of science was ever made as the result of a generally-expressed need for that discovery, or invention in answer to universal demand. Inventions come because they are possible, not because they are wanted, and scientific inventions come in precisely the same fashion; and in the

SCIENCE WILL PROVIDE

case of every invention its ulterior effects are what nobody has wanted and most people would acutely dread. Science as a whole will surely go on to the discovery of things we little dream of; and the effects of those discoveries will be such changes in human life and civilization as we can hardly tolerate to think about. To bring about one of these inventions in answer to our prayers is only just short of impossible. To check the general flood is equally impossible. In all the affairs of men, science included, the wind bloweth where it listeth.

CHAPTER XXI

THE GLORY THAT IS ART

A PERIOD of enlightenment is always known not by its inward causes but by its outward product. When we think of the glory that was Greece, or more specifically Athens, we do not dwell upon the naval hegemony by which Athens profited so largely and from which it was able to divert the funds to build the Acropolis and to indulge in those theatrical performances which have been the admiration and despair of poets ever since; we turn instead directly to Scopas and Euripedes. These, we say, are the men who created all that glory. The Renaissance glows incandescent in our imaginations not by reason of trade routes, or the mercantile marine of the Italian cities, or the autocracy of the Holy Roman Church, but rather for the fine flower of its painters. Our own age is no exception to this rule. We have with us to-day machine technology and capitalist imperialism. But we are neither fully aware nor altogether proud of these expedients. What Ruskin and Morris so bitterly condemned as ugly, very few of us have had the courage to look squarely in the face. From the point of view of culture, machinery is still on the whole deplorable and trade a necessary evil. The proudest possession of the twentieth century, our one superb and undoubted cultural achievement, is modern science.

As a mark of enlightenment, science runs true to form. Art, poetry and metaphysics are the marks of culture not only because they are the decorations and refinements of the higher life, but because as decorations they are capable of perfection. Civilization as a whole, the common life of manufacture and make-shift, is too inchoate to be esthetically appreciated. What is perfection of form in the life of peoples? Our reply to this question is always an evasion. We can not applaud a people. Therefore we select whatever of theirs is most compact, most precise, most independent or detachable—their churches, their mural decorations, their poetry, philosophy, or learning. Our admiration of science may not seem to refer to its esthetic perfection of form. We may, of course, think of it in purely utilitarian terms, honoring it only as the designer of our machinery. But in that case, we are not making it the basis of our claim to live in an era of enlightenment. If such is our boast, if our feeling for science is more than utilitarian we will still be only slightly aware of the esthetic principles behind our blandishment. But that is also the case with all forms of art. Only critics translate their esthetic emotions into the language of "significant form." Others feel that what painting or poetry has revealed to them is the soul (or mind) of a people. What they have obtained from listening

SCIENCE: THE FALSE MESSIAH

to music or viewing architecture is a glimpse of higher truth. This is precisely what we feel in the case of science. When we are contemplating the non-Euclidean geometries, or the quantum theory, or bacteriophage, we receive the distinct impression that we are standing in the presence of deeper and broader truths than any people has ever before attained.

If we fail to qualify—"has ever before attained in just this manner"—that too is natural. Each people, in each enlightenment, feels that its manner is the only important one. The Pyramids pale before the Parthenon; the Parthenon crumbles to inanimate dust before La Gioconda; La Gioconda stands mute before the Dark Lady of the Sonnets—and so on. And as we survey the panorama of science, spread out between the spiral nebulæ millions of light years distant, and the magnitude of the electron, we may be pardoned for the slackening of our enthusiasm for other forms of art, for our devotion to science as the profoundest revelation of the mind of man, and for our own as the century of greatest enlightenment to date. If the Renaissance was incandescent, the present is ultra-violet! Where is there any work of art more perfect in form than Newton's laws of motion—unless it be Einstein's theory of relativity?

Consider the riches of our gallery. We are now aware, for example, that human destiny depends entirely upon various hereditary qualities, transmitted by parents to their offspring. Consequently,

THE GLORY THAT IS ART

we know that to improve human society we have only to control the conditions of mating by a program of eugenics. We are also aware that human character is shaped by climatic conditions, and that the hope of the future lies in the manifest destiny of the inhabitants of temperate zones, the mountain dwellers, the seaboard peoples, and those who are exposed to extremes of heat and cold. We know further that the regulating condition which governs human growth and development is diet; that size, stamina, intelligence, and the possession of special faculties result from the eating of iodides, which are accordingly the key to the millennium. Moreover, we have become convinced that the secrets of human excellence are locked within the ductless glands, that failure and success are functions of hyperthyroidism, and hypopituitarism, and can be corrected and controlled by gland transplantation. In addition to all these things, we are indebted to the psychiatrists for the discovery that human destiny bears little or no relation to the physical organism but is a matter of behavior, especially of infantile sexual behavior, through which the ruling complexes are established; and that it can be controlled by psycho-analysis. We live in a period rich in information.

In view of all this, it is not surprising that we feel the beating of the invisible wings of imminent perfection, and hasten to prepare ourselves for the translation. The multiplicity of prophets and straight and narrow paths does not dampen our

SCIENCE: THE FALSE MESSIAH

ardor; it actually heightens the intensity of our faith. What is most significant in the modern enlightenment is the universality of the conviction of millennium. One man expects a Messiah in the likeness of Herbert Spencer, whose brow was so smooth because, as he said, he never was puzzled! Another watches and prays for the reincarnation of Binet, or Francis Galton. But they concur absolutely in their expectation of salvation, and in spite of all discrepancies in their philosophies—which may proceed in exactly opposite directions—they actually reinforce each other's faith, each rejoicing in the "fine optimism" of the other just as Methodists rejoice in the discovery that Todas and Tlingits also worship the one true God. All things work together for good to them who love God, and all paths lead to perfection, however divergent they may appear to the eye of carping criticism!

The sheer cumulative multiplicity of intellectual hypothesis which the scientific enlightenment has presented to this generation is responsible for a depth and breadth of optimism unexampled in past history. Every age has its optimism. Hence the biological law: hope springs eternal in the human breast. But apart from metabolic causes, optimism seems to be an accompaniment of intellectual effort. Every thought is something of a guess, and every guess is a hope and expectation of fulfillment. To make an invention without expecting it to work, or to record a discovery without assuming that it will be verified, is humanly impossible; no one has ever been so detached except possibly the White Knight. And from the expectation of success, it is only a step to the belief that each new discovery is epochal. No doubt the first man to cultivate wild grass and the first man to chip flint both imagined that their innovations had solved all the fundamental problems of mankind and that no serious difficulty ever need arise again to plague posterity—if only those achievements could be brought at once into general employment.

Modern innovators have always thought so. The story of utopias is the story of imagination plus fatuity. Condillac's theory of infinite perfectibility was the obvious and inevitable corollary of Voltaire's belief in reason, of Montesquieu's belief in law, of Rousseau's belief in democracy and education. Malthus played devil's advocate to Condillac, writing his essay on the law of population to expose the futility of such aspirations. But he fell a victim to his own ingenuity, as even a devil's advocate must do, and discovered a way out of the clutches of his iron law by the route of "intelligent voluntary restraint" of reproduction—that is, by the immediate application of his epochal discovery. The application did not transpire; but a reader of the name of Darwin was able to show that even the iron law naturally selected those fittest to survive, to which Spencer immediately added the obvious inference that societies no less than individuals move steadily up the ascending scale from the

SCIENCE: THE FALSE MESSIAH

simple and debased to the complex and illustrious. Thus do we drive even war and pestilence in double harness toward the universal goal!

Indifference to history is the essence of optimism—particularly indifference to the history of optimism. It is impossible to study the history of a religion and retain an undiminished faith in the tenets of its creed. The same is true of the history of theory. Thus, for example, the faith of our enlightenment derives in large part from the discovery of "society." Ostensibly the "science of society" is a compendium of social laws as sound and sure as those of physics, and is therefore in a position to control "social development" just as science (ostensibly) controls nature. Actually, every one knows that neither nature nor social development is under any sort of control at the present moment. Actually, our glad tidings are the expectation of control in the future. We rejoice in the tidings just as people rejoice in being cured by faith of the "fear of" cancer. The thing itself is present only in imagination; and imagination is fed not on evidence of accomplishment but on the assumption that such things must be. They must be because we see that they need to be. Science discovers nature, and straightway we see that nature sorely needs to be controlled. Social science discovers society, and we see that to control it is a deplorably obvious necessity. Thus the thought is father to the wish.

A strict calculation of chances reveals, however,

that the discovery of a condition may have the effect (1) of improving it, or (2) of aggravating it, or (3) of leaving it precisely where it was before. We assume that our present self-consciousness about the workings of society must certainly enable us to deal with those workings in short order. This is "telesis" to use a word coined by the great American sociologist, Lester F. Ward, to fit this very case. Telesis is the conscious guidance of social development toward a previsioned end. Genesis is "mere growth." Human history is divided into two parts: one, from the beginning of time to the advent of sociology, the genetic part, wherein growth took place by natural process unintelligently; the second, beginning with Ward, or at least with Spencer, or at the earliest with Comte, is "to be" telic—intelligently directed to the end of ultimate perfection. We can surely preconceive this end. It is, merely, all that we now are not. Seeing that we now are the reverse of what we are not, it seems to many hopeful sociologists quite incredible that we should fail to take advantage of the opportunity of becoming the reverse of what we are, much less that we might make a mess of trying. Yet if God intended the important things to be a secret from His human children, another Eden fiasco as a consequence of more ill-gotten information is a distinctly suggested possibility.

There is also the possibility that the discovery of society, the discovery of heredity, the discovery of climate, and topography, and glands, and com-

SCIENCE: THE FALSE MESSIAH

plexes, leaves these things very much where they were before. Is there any evidence to the contrary? Human history has been one long, uninterrupted series of problems and crises giving way to more problems and crises to be followed by further problems and crises. History is a continuous succession of solutions which did not solve and expedients which failed to expedite. Has this history shown any notable alteration since the death of Ward in 1913? It would not seem so. Expectation of utopia is not founded upon history.

But there is a way to explain all this. Genius, we say, is always in advance of its generation. Consider Plato. His *Republic* is surely one of the finest works of the human intellect. What future-piercing suggestions does it not contain for the orderly arrangement of the organs of society! Breeding by the physically fit! Government by the intellectually potent! Equality of the sexes! Education over a period of forty years for the performance of the highest services! Every one knows that none of Plato's millennial suggestions has ever been put into force without crippling reservations; but no one has ever ventured to conclude that this means that the *Republic* might as well never have been written. On the contrary, its neglect is a convincing proof that the intellect is mighty and able to prevail: ever since Plato's time we have possessed the means of founding the perfect state! No such state having been founded, our confidence in its possibility remains unshaken.

But the perfection of the *Republic* is the perfection of artistic form. There is a saying current in the normal schools that if the pupil fails to learn, the teacher fails to teach. Why should this not be true of kings and legislatures, of electorates and cultures? No doubt it should. If a nation fails to act, a prophet fails to stir. Yet if he has stirred his hearers or his readers he may have done his part. He has created what is, in its perfection of form, the perfection of the Republic, and the Sistine Chapel, and the Periodic Table of the Elements (why should we not use capitals in chemistry?), a supremely stirring work of art. Civilization is what it is and art is what it is, and ne'er the twain shall meet. The body politic proceeds by expedients of its own, waxing great or little as the case may be. The master minds solve all the problems to the complete esthetic satisfaction of themselves and all their readers; and if they have to wait upon posterity for the adoption of their proposals, who shall say that they are not in advance of their generations, and farther in advance the longer they have had to wait—upon the shelf! The very thought casts aspersion upon Plato! It is enough for works of art that they achieve immortal distinction as works of art. Let the dead bury the dead and machines fabricate machines. Artists dream of mathematical perfection and social order. Golden dreams! The darlings of our national academies! But are we also to be improved by Napier's Analogies? by the Mendelian Laws? by the Libido? Are we to be improved by the *Ode to a Skylark?* It is not necessary!

CHAPTER XXII

THE SCIENTIFIC MIND

IN THE Republic of Letters, no dogma is more precious than the inexhaustibility of knowledge. We would fain believe that all can know, all can see, all can appreciate. This notion—it is only a notion—is based not upon observation, but upon logic. As a matter of logic it does appear that a work of art is not diminished by admiration. Unlike riches, its bounty can be indefinitely shared. So also with knowledge: if my neighbor learns, that does not make me foolish. He does not learn at my expense, nor at the expense of any one. Whatever he is able to know is a clear addition to the total of learning in the world. As a matter of logic, therefore, it seems reasonable to discuss the possibility of all the people in the world being able to appreciate Picasso and Stravinsky, to use the higher mathematics, and to regard all things in the light of science.

Logic aside, however, there are great practical difficulties. To attain such altitudes of culture requires years of application to that end alone. This is why so very few achieve the highest summits and no one absolute cultural perfection. As an aspect of our culture, science is no exception to this rule.

THE SCIENTIFIC MIND

Scientists frequently argue that apart from making science one's profession, the chief advantage to be obtained from the study of science is the scientific mind: freedom from dogma, hospitality to unexpected truth, the experimental attitude. Perhaps this is true. Artists often say the same of art. If one is not to be a painter or even a critic, one may still gain something in fineness of perception, sensitiveness to form and color, awareness of the beautiful, by cultivating an acquaintance with good paintings. Theoretically, all may gain these insights. But actually, only a few of us have ever done so. Theoretically, we might become a scientific people; but we have not, and are not likely to, except in the sense in which we are now a Christian people.

It is too bad that universities can not wax great in the same geometrical proportion by which they can increase. It is too bad that epochal discoveries turn out to be discoveries of problems and not of the means of solving them. In particular it is regrettable that the world at large never quite develops the critical mind, the experimental attitude. But a disappointment so constant must have a natural cause. Intellectuals may take whatever comfort they can from the reflection that the persistent stupidity of their contemporaries is not due to any weakness of theirs but is the natural and probably unchangeable condition of large masses of people.

When civilization has been defined as an

SCIENCE: THE FALSE MESSIAH

accumulation of habits, the experimental attitude is automatically excluded. There has been some talk recently of the "habit of intelligence"; but the phrase is an Irish bull. No doubt this makes it exciting. To plan a course of study that will train people to be experimental minded is an exhilarating adventure in pedagogics, just as squaring the circle use to be in mathematics. The circle and the square are the twin poles of geometry and ought to stand in some regular and even relation to each other; and no doubt they would if the universe had been constructed by a creator with a sense of harmony, or even of humor. But in fact that fascinating ratio is an unending decimal—3.14159 and so on—and in fact "the habit of not forming habits" is just so many words, interesting and piquant words, as Irish bulls so often are, but not on that account very helpful for the education of the young or the inspiration of the old.

Nevertheless, notwithstanding the paradox, the habit of intelligence is baited with possibility. Habits in the ordinary sense are not like eating: within limits they can be dispensed with, if in no other way, at least by doing nothing. Of course, any one who carries his distaste for habits to that point will require to be fed and otherwise stabled like a helpless animal; and this means that such experiments can be indulged only occasionally and by individuals who are surrounded by a sufficiently compliant retinue of helpers. For the population at large self-help is always a necessity, disagreeable, perhaps, but unavoidable. Furthermore, human kind being unable to subsist by instinct and in isolation, self-help requires the unwitting co-operation of a community—in short, organization. This organization imposes habits on the individual and institutions on the community; otherwise the crowd is not an effective community at all. Thus it appears that while a human organism may exist, experimentally, without extensive habits, or any sustained activity, human beings can provide themselves with the means of subsistence on no other basis except habit. Certainly they have never done so, nor do the modern intellectuals intend that the practise shall be started now. On the contrary they dote on sociality!

Here is the seat of the paradox. The basis of sociality is institutions, and the basis of institutions is habits; and habits, formulas, dogmas, and institutional rigidities variously are the antithesis of intelligence and the nemesis of the experimental attitude. That is why education runs perpetually from one dogma to another, making a formula of every inspiration—even that of the habit of intelligence!—and producing a thick growth of academic weeds out of the fertile soil of original ideas. The essence of education is tradition because education is a joint enterprise and tradition is the essence of all joint enterprises. From this inexorable logic there is no escape.

Except for individuals! Thinking is an individual enterprise. One can indulge it even in bed, as G. K. Chesterton once remarked. Consequently

SCIENCE: THE FALSE MESSIAH

one can do it without working, and in a hermitage. No doubt even this is a ticket-of-leave freedom. People who have grown up in domestication are not going to shake off all the "shackles of civilization." For all his splendid isolation your hermit shows most of the marks of caste, and is a good Bible-reading, Gregorian calendar-keeping Crusoe, treading the paths of a middle-class Englishman through the tropical jungles of his south sea isle. Nevertheless he can be a maverick in many ways. Robinson Crusoe might have taken many more liberties than he actually did if he had not held constantly before him the ideal so to live that he might become suitable reading matter for the tenderest children and a twin figure upon the shelves of Sunday School libraries of the pious Swiss Family Robinson. From his sinecure he might have flouted Parliament with impunity. It would have been possible for him to hold opinions disapproving of the recent war. He might have applauded the acts of Lenin. He might even have read the works of Sherwood Anderson (taboo in France) and of James Joyce (taboo in America). He could have discussed birth control with his parrot, and written letters to Man Friday attacking the monogamous family system. He could have spent his Sundays chanting the Devil's Mass and his Mondays scrubbing his cavern floor with the Union Jack. He might even have come to lunch without a collar and put his elbows on the table or committed the unforgivable sin in the matter of chicken salad. In short, he had a great deal of leeway which he failed to use.

THE SCIENTIFIC MIND

Every one who cares to endure isolation from his fellow men can do all these things and much naughtier ones if he can think of any. Indeed, individual genius is precisely the capacity for doing so—not of course for performing the absurdities catalogued above, but for walking in untrodden fields. When intellectuals talk to us about freeing our minds of the idols of the theater and marketplace, they mean that we are to look upon the religion of our fathers and the patriotism of our friends with a cold inquiring eye; to try all things and hold fast only that which is good. They have done this, in various degrees, and have found the result excellent—for their minds. It has improved the clarity of their observation and the precision of their thinking. All of which is an unqualified advantage—from the point of view of the intellectual life.

But life is not all intellectual life. On the contrary, most of it is extremely practical and common drudgery; and from this point of view the intellectuals, fine flowers though they be, are parasites clinging to the bark of civilization. Thus while the critical mind flourishes upon the rare essence of individualism, the drudging mind sends down roots into the coarse and fetid soil of common tradition, standardized, institutionalized, accumulated from the droppings of countless generations of similarly stupid, obvious, and standardized Children of the Lord. From these mews opium offers a temporary and intelligence a permanent escape. That, as all philosophers have said, is its power and its charm.

SCIENCE: THE FALSE MESSIAH

The mind thrives upon negatives. Critical and tentative, it frees itself from prevailing obsessions. Turning the inner eye from immediate and necessary things, it contemplates perfection and infinity. It sits by the side of the road and lets the world pass by. With persistent curiosity it observes the spectacle of men and nations and makes amused and caustic comments. From the point of view of philosophical detachment, nearly everything in the world is done wrong. To limpid intelligence the world is a mud-bank of deceit on which crawls the human race gorging itself on self-deceit.

This point of view is not manly; it is presumptuous. To err is human, we are told; to review, divine. In his famous essay on Utilitarianism, John Stuart Mill laid down the principle that no man who has ever experienced both lower and higher pleasures has preferred the lower. What "lower" and "higher" mean in this discourse has been much debated and can never be settled now except by saying that Mill thought it right to call experiences men will not give up the "higher" ones. Farther than this we can not go. But this is a very great way indeed. The tenacity with which men who have gained intellectual insight hold on to it, sacrificing if need be all the other goods of life in order to retain this one (as they think) supreme endowment of seeing life steadily and seeing as much of it as the eye can reach, this is an inspiring and moving thing before which even the commonest of men will pause with awe and something akin to reverence. Man is most god-like in his mind—at least in this sense, that the highest god has always been conceived in the image not of the banker, the athlete, or the politician, but of the sage. Man climbs farthest above himself in the intellectual life; and science no less than music, painting, or philosophy, is a ladder upon which he climbs. Whether or no such heights are vertiginous is another question. Perhaps they are. Professor Warner Fite, a philosopher, has warned us that "men who make sophistication their profession must not expect repose." Perhaps this, too, is another cause of the unfitness of the intellectually superior.

In the intellectual life—in science as well as in philosophy—man does climb above himself. These awe-inspiring heights are not the appointed habitat of what is always and forever an animal species. Not only human nature—all nature lives on another level upon which the race of men, including intellectuals, is physically nurtured. This is the level upon which civilization sprawls and spawns, a lowly thing made of the dust of centuries. Yet for all this, civilization also is like Jehovah: it is that it is.

CHAPTER XXIII

THE RULE OF PSEUDO-SCIENCE

ACCORDING to the great author of *Folkways*, those social customs, upon which alone civilization rests, survive and are passed on only because they have proved their usefulness. What Sumner meant by this is doubtful. If he meant that all things work together for good, we are obliged to shrug our shoulders at finding such an unscientific idea in such an earnestly objective book and to explain it as a relic surviving in Sumner's own little universe of folkways from his days of preacherdom. But usefulness may mean many things. From the point of view of chemical analysis, the miracle of transubstantiation is not particularly useful. Yet that sacrament, together with others, unscientific though they may have been, may nevertheless have served the purpose of holding the European peoples together under a single roof of folk-lore; and from the point of view of their descendants that seems a very useful thing indeed.

From this point of view, Sumner's dogma appears to be a truism and therefore most scientific. It proclaims that no folkway can survive except by performing some function which leads to its sur-

[282]

vival; and this function must be called useful not according to modern standards but according to the actual situation in which the folkway exercised its sway. Freethinkers like Thomas Paine never encounter the slightest difficulty to proving that the religious beliefs and ceremonies of all the peoples of earth are a mere encumbrance of good fertile soil—as regarded in the light of reason and from the security of the eighteenth century or any other age of teeming reasons. But could those half-starved barbarians who slaughtered their neighbors and even each other so outrageously under the command of Moses and Joshua ever have reached their promised land and handed their preposterous superstitions on to be a pillar of light to the Noble Nordic, had they applied the tests of Paine to their much-advertised stone tablets? The usefulness of those tables was, as we say, sacerdotal; but who will say that such things are of no account?

The folk-lore which has sprung up about us in such luxuriant profusion during these recent years can also be subjected to the tests of Paine. When those tests are made, their results indicate very strongly that science is not "scientifically" qualified to be a guide to our civilization or any other. To be sure, it has made pretensions. But so does every folk-lore. The fact of pretensions is no evidence of qualification for anything but the sacerdotal function of making pretensions and getting them believed. Moreover, the pretensions of science arose not from any discoveries within the

SCIENCE: THE FALSE MESSIAH

field of science but from the embarrassing position in which the men of science have been placed. They have been charged with weakening, even perhaps demolishing, our pre-scientific, Christian folk-lore. As none knows better than they, such was not their original program. Originally they had no program. Originally they were only a parcel of curious and inventive people who were interested in scheming out a set of theories to accompany the mechanical inventions of their day. That mechanical invention flourished mightily was not due to their eloquent persuasion but rather to a peculiarity of mechanical invention itself. People think devices harmless. They employ them. They encourage the contrivance of more of the same. Thus they water the soil from which science springs. The soil was watered so much more effectively than any one expected that scientists soon began to be embarrassed at the extensiveness of their own theories. They devoted almost incredible ingenuity to disavowing evil intentions toward the older folk-lore, to bringing the assistance of scientific proofs to bolster the older folk-lore up, to denying even the possibility of a conflict between science and religion, meaning other religion. But the tide was set against them. This age was fated to become mechanical, and science was fated to ascend the throne of highest possible prestige. It is to science that we now look for social guidance and for spiritual consolation—not because it is reasonable to expect those blessings from the law of gravitation any more than from the stone

THE RULE OF PSEUDO-SCIENCE

tablets of Sinai, but because our eyes are blinded by the glory thereof.

Unless all signs fail, scientists—or persons who will be called scientists—are to be the rulers of our future. How this is to come about remains to be seen and will anyhow be described in differing terms by different writers even after it has occurred. One obvious way to put it is that scientists will gradually supplant our present rulers in commerce and industry and the lesser affairs of political government. To depict this scene we have only to unfold the panorama of modern life. Wherever we look, our affairs are becoming not only more complex and involved than ever heretofore, but especially more mechanical. The machinery by which life is conducted more and more staggers the imagination of all of us who are not "experts." In this situation the employment of experts is more and more an urgent necessity, and dependence on the experts constantly increases. The end of this process is clearly the final dominance of all our affairs by men of scientific training.

The same transition can of course be depicted in more dramatic colors. We are in for revolution. It might seem, perhaps, that a Christianity which has proved flexible enough to adapt itself to the mammonry of modern capitalism would have little further trouble with a mere intellectual detail of capitalist civilization such as machinery—or science. But against this complacency, shrewd revolutionists will argue that the fundamental fact

SCIENCE: THE FALSE MESSIAH

is that science and invention are not a detail of the prevailing mammonry. On the contrary, the real break occurs just at this point. It was much less of an adaptation than we supposed for Christianity to embrace capitalism—and that just because capitalism is much more closely related to the feudalism of the past than to the "industrial civilization" of the future. Capitalism is the last expiring transformation of feudalism, just as modernism is the last expiring form of Christianity. The relation between the two is close and necessary. The very anxiety of the pillars of contemporary mammonry to maintain our ancient Christian piety shows the atavistic character of capitalism itself. We have here two aspects of the same folk-lore: one descending by "reformations" from the medieval church, the other developing by legal fictions and interpretation from the feudal rights of a social hierarchy based on the ownership of property. In contrast to these, industrial society will be organized not about ownership but about expert direction; and just as historic Christianity served to sanctify the feudal order (although such was not the urge that gave it birth) so the folk-lore of the future will be a scheme of ceremonies and suppositions favorable to the exalted position of the experts.

Not that revolution is brought about merely by literary contrast: upheavals follow social pressure, and social pressure is raised only by the fires of revolt against oppression. But oppression we have always with us. In each civilization it takes its ap-

THE RULE OF PSEUDO-SCIENCE

propriate form, being known among us as the condition of the industrial proletariat. When and where and specifically why outbursts occur, only the most doctrinaire of revolutionists pretend to know. In any case such details are incidental if not accidental. In periods of expansion, mammonry becomes arrogant, and after that—the deluge. Mr. Brooks Adams ventured to make a scientific law of his *Theory of Social Revolutions*. The more obsolescent a social order becomes, so he said, the more rigid it is. The more rigid it is, the less it is able to give way to the pressure of social necessity and the demands of the times. The less it can concede to the demands of the increasingly disgruntled classes, the more violent the ensuing explosion is bound to be. In our own case, it is well known that the industrial population is being fed on a gospel of scientific rationalism, mechanical causation, technical efficiency, and the rule of the expert—of course "in the interest of society." Pentecostal expectations and humanitarian zeal are the constant features of these revolutionary gospels. What is noteworthy in the evangelism of the modern proletariat is the unmistakable drift away from Christianity as tainted through and through with mammonry and toward science as the means of realizing the revolutionary dream.

Even here we must guard against the shallow intellectual theory of revolution as a product of "parlor Bolshevism." The upheaval is never in any social crisis brought on by the theories which decorate it. On the contrary, the theories are brought out by the events. Mr. Veblen has pointed out in various books that the obsession of revolutionary theorists with mechanics is induced by the habituation of the industrial under-dog to machinery. The proletariat is an industrial proletariat, a factory proletariat, accustomed above all things to mechanical technology. Inevitably, therefore, technology supplies the pigment with which the revolutionary dream is colored.

This does not mean that the proletarian dream or the revolutionary theories will be realized. Such dreams never are. It is an axiom of revolution that, so far as social stratification and perhaps social injustice go, violent upheavals invariably leave us very much where we were before. One set of rulers is unseated and another promptly takes their place. Whether this fact too can be stated as a scientific law is not quite clear. A perusal of the history of these affairs suggests that the clearing of the dust of turmoil usually reveals that class which was formerly just below the former ruling class now blandly occupying the seats of the mighty. If this is true, it is something of a paradox. Any class of men who are so near to power as such rulers-elect must be loyal supporters of the old régime. They must feel, throughout the events leading to explosion, that their interests are upon the whole with the established order. This, for example, appears to have been the attitude of the substantial citizens who formed a majority of the Long Parliament.

THE RULE OF PSEUDO-SCIENCE

Though far from being Royalists, this group—the nucleus of the upper middle class—were nevertheless Loyalists, and had to be prorogued by Cromwell before revolution could be consummated. But after the throes of revolution had passed, there seemed to be nothing to do but recall the substantial citizens of the Long Parliament, who proceeded to rule as by established right, wiping out the work of intervening men, and restoring as nearly as possible a *status quo ante,* purged only of Stuart royalist pretension.

The necessity for such a restoration seems to be that, after all, only the substantial citizens are capable of ruling. And as the "substantial citizen" is defined by the situation in which he appears, the meaning of the social law seems to be this: that the transfer of authority is determined not by the violence of the classes which upheave but by the nature of the problem civilization has been all the while presenting. The problem being in the main industrial, it follows by social necessity that the real power must pass to that class of men best qualified to grasp and hold it in industrial society. This means experts. Feelings and attitudes toward the old régime, or toward revolutionary violence, have little or nothing to do with this logic. The ascendency of technical experts under the established capitalist system is due to the demand of that system for technical control. The same fact naturally brings the expert and technician into close association and sympathy, or at least a sense of community

SCIENCE: THE FALSE MESSIAH

of interest, with the powers that be. As any one can see, few indeed are the experts who venture now the slightest word of criticism of the prevailing mammonry, or who seem to wish to do so. Nevertheless, by the irony of fate or social law, this is the very class which is almost certain to profit most by social revolution. The reluctant recall by the Russian Bolsheviki of the "bourgeois" technical experts is a most significant phenomenon.

But another and more important irony of revolution is the effect of power upon a newly established ruling class. Whatever the manner in which the conquest of civilization by science takes place, we hope and pray at least for a golden age of science. Under the rule of scientific specialists, science and invention should have the freedom of the land. But this may or may not be the case. Under the rule of capitalists, business should be quite free and encouraged on all sides. Nevertheless, some enterprisers have a way of usurping disproportionate encouragement while others struggle against all sorts of subtle obstacles. The reason for this partiality is that whatever men are in positions of authority are not only members of a class with the feeling of a class but individual members of that class. As individuals they have much more specific interests at stake than the vague requirements of class loyalty. As every one knows who has had any experience of universities or research foundations, the same is true of scientists. While no scientist is insensitive to the case of science as a whole, individual scien-

THE RULE OF PSEUDO-SCIENCE

tists are frequently—perhaps usually—more taken up with one research than with another; and often this partiality goes as far as favoring one branch at the expense, even at the total expense, of another. It would be pleasant if we could affirm that men put such petty personal concerns behind them when they assume the responsibilities of power; but we can not. Alas, it is not true!

Add to this the fact that whenever power is in the offing it is the men of administrative bent who lay it by the heel. Not even revolution is likely to alter this universal social law! Under the dispensation of industrial society we may expect the ruling class to be not owners but experts, technicians, engineers. Still, administration will continue to be administration. Responsibility will seek out not the brilliantly original, not the gentle dreamers nor the wildly imaginative schemers, but the solid, respectable, thoroughly well-versed master mechanics of the prevailing order of technology, the men who understand how things are now conducted and are able to inspire confidence in their ability to conduct them well along established lines. They are the men—vary the setting as you will—who conduct every civilization. Such men are notoriously inhospitable to innovation. They are sober men, pious men, science-fearing men, who mean to stand by the sacred traditions of their creed. Many critics have pointed out what happened to Christianity in the early centuries of this era when it ceased to be a religion of revolt and social protest and became a religion of empire and conquest, of popes and bishops and a universal church. We have no reason to expect otherwise of science. When the pillars of society are the priests of science, will freedom of thought in science be cheaper or dearer than it is at present?

In this age of enlightenment we are very proud of the tolerance we show religious sceptics—that is, persons who doubt the Christian dogmas. Does this mean that tolerance has increased in all men's hearts or does it mean that Christianity is no longer the center of our jealous interest? Before answering this question we must note that we do distinguish between dogma and the truly religious spirit. At this moment of transition we still are able to distinguish between the truly scientific spirit and the arrogant pronouncements of the self-appointed priests of science. But shall we always be able to do so? Shall we be able to do so when these pundits are the accepted rulers of society? It would be strange indeed if the final proclamation of science as the guide and protector of mankind should ring in as well a day of scientific inquisition—a day when to doubt the word of Newton would cause any rashly heretical Einstein his excommunication from the universal church. If such should prove to be the case, the philosophers of that distant age would look back upon the present era of scientific enlightenment with a smile. They would see us feeding our imaginations upon a wraith of hope—as peoples wandering in the wilderness of transition have

THE RULE OF PSEUDO-SCIENCE

always done—that the coming of machinery might mean the translation of mankind. They would see that machines had come indeed with force enough to disestablish the most ancient and rigid dogmas. They would see the Pentecost of science, and the charging of the disciples of the new evangelism: go ye unto all the world. But after this they would see the gradual emergence of the dogmas of the new tradition—the popery of pseudo-science.

IN CONCLUSION

THESES TO BE NAILED TO THE LABORATORY DOOR:

1. That the truth of science is established only by belief, after the manner of all folk-lores.
2. That scientific formulas, however charmingly mysterious they may be, do not touch the central problems of living.
3. That the credit of science rests wholly upon its connection with machine technology, of which it is a part.
4. That machine technology was not derived from science, but crept upon us after a fashion of its own.
5. That there can be such a thing as too much machinery.
6. That our industrial revolution is located not in the past but mostly in the future.
7. That the vaunted freedom of modern civilization is in fact the loosening of the bonds of order and belief by industrial revolution.
8. That the dissolution of our institutions has gone farther than most of us suspect.
9. That the wane of our ancient culture is due not to the influence of scientific ideas upon our minds but to the effect of machinery upon our lives.

IN CONCLUSION

10. That scientists have always deprecated the supposed effects of their discoveries—fruitlessly, since other and still more disturbing discoveries have always followed.

11. That by trying to make our beliefs scientific we have succeeded only in making them absurd.

12. That we can keep science and belief separate by relegating our religion to the Sabbath Day.

13. That although the highest truths of religion can be proved, they lose in the process all meaning and value to humanity.

14. That creeds and churches are reformed only at the expense of losing all their power.

15. That the scientists who proclaim that science bolsters up religion are deceiving us and possibly themselves.

16. That the laws of science are not statutes; they are definitions.

17. That the divine order, which we must "let alone," was discovered not only by science but by special interest.

18. That if we examine our lives and our civilization in the light of science, we see only that they are a natural growth.

19. That the facts of science can be translated into the guidance of human destiny only by astrologers.

20. That inventions are provided not to suit the needs of civilization but according to the development of science and invention.

SCIENCE: THE FALSE MESSIAH

21. That the perfection of science is a perfection of form, and our glory in it the pride all peoples feel in their great art works.

22. That the minds of a few men can be sublimely elevated by the study of science only if the minds of most men are regulated by tradition to humbler but more necessary ends.

23. That when science has become supreme any attempt to rectify its formulas will be persecuted as heresy.

THE END

HOLIER THAN THOU

THE WAY OF THE RIGHTEOUS

The more I look the facts in the face . . . I just know there's another woman mixed up in this affair. Tell me! If it wasn't true why would I find it so easy to believe? There's proof right there!

—Mrs. Bungle.

Ich musste also das Wissen aufheben, um zum Glauben Platz zu bekommen. . . .

—Immanuel Kant.

(I had therefore to remove knowledge, in order to make room for belief.

—Trans. Max Müller.)

CONTENTS

CHAPTER		PAGE
I	PRE-VIEW	11
II	MAN THE ANIMAL	20
III	THE BASIS OF DECORUM	30
IV	O TEMPORA! O MORES!	45
V	GOOD TASTE	63
VI	FINE FEELINGS	79
VII	GOODNESS SANS SUPERSTITION	93
VIII	PIE IN THE SKY	108
IX	MORALS FOLLOW THE FLAG	123
X	GIDDY EMINENCE	137
XI	ONWARD AND UPWARD	155
XII	CONTROL BY EMULATION	178
XIII	THE RIFT IN RIGHTEOUSNESS	195
XIV	LAW VERSUS ORDER	127
XV	NO MORE CRUSADES	236

HOLIER THAN THOU

CHAPTER I

PRE-VIEW

LEADING a pure and noble life is precisely the same kind of thing as dressing properly and taking off one's hat to a lady. Righteousness, good manners, fashion—they are all one. The sole compelling force behind all morality is the public opinion of any given time and place; and the sole motive of every decent act is one's preoccupation with what the neighbors think.

Neighbors vary, of course. No one supposes that the neighbors of Mr. Upton Sinclair, the author of *Oil,* and Mr. Harry Sinclair, the author of the late lamented lease to Teapot Dome and the equally lamented Continental Trading Company, Limited, are precisely the same neighbors. The discrepancy may even be considerable and far-reaching. In a quite legitimate sense Mr. Upton Sinclair is a neighbor of Lenin and Karl Marx; while Mr. Harry Sinclair is clearly a member of that gay company which includes, among many others, Jim Fisk and the redoubtable Captain Kidd.

If we are to be serious about the substantial identity of styles in righteousness and styles in all other kinds of habiliment—and it is the intention of this book to be quite serious—we are bound to take account of two more or less distinct respects in which all styles correspond: their cause and their effect. An illustration will exhibit what these are. In one of the most delightful passages of *The Theory of the Leisure Class* Mr. Veblen has sketched an amusing hypothesis of the cause of the rapid alternation of the styles. The idea is that each style comes into being as a revolt against the intolerable ugliness of the preceding style. Yet each style, though it begins with a determination to be different, is ruled in its development by the same principles which have guided all its predecessors. Like them, it must express the canons of invidious display and conspicuous waste; and since these are not esthetic canons by any means, the effect of their guidance is a fresh atrocity of taste, no more tolerable than any earlier example, and therefore just as transient. But this theory, ingenious as it is, does marked violence to many of Mr. Veblen's own theories. In the first place, as he insists—and we must perforce agree—there is no standard of absolute good taste, esthetic or otherwise, to which a really satisfactory style might approximate. On the contrary, taste itself is formed by the same invidious canons which rule the styles. All styles are admired during their brief heyday. In the second place, it is not necessary, under Mr. Veblen's own theory, to invoke a revolt of taste to account for the alternation of the styles. That is sufficiently motivated by the very canons of conspicuous display and invidious waste. What brings about the continuous uneasy shifting of the styles is the necessity on the part of the rich to keep two jumps ahead of the poor. The rapidity of the succession depends wholly on the readiness, in a given social situation, with which the poor can ape the raiment (and other social gestures) of the rich. A garment is out-moded when it has been copied in the "trade."

If we inquire what causes any given style, we are faced with a problem which, though difficult, is wholly secular. Doubtless no two styles proceed from the same specific causes. Each is a product of a unique situation. But the ingredients are always similarly unpretentious. The present vogue of the "revealed knee," as it is called in the advertising of hosiery concerns, is not derived from any supposititious deference to any hypothetical ideal of beauty. Its ingredients are probably of this kind: the present sexual freedom, of which the revealed knee is one expression; the present popularity of sports for women, which has necessarily led to the recognition that females have two legs; the great extension of women's freedom of movement in every department of life, of which the foregoing are simply special aspects; and behind all, the mechanical revolution in our way of life which has made cloistered womanhood a physical impossibility. The revealed knee is, in a definite sense, an expression of the machine age.

Much the same thing will be true of the particular styles—in clothes and in morals—that flourish in every given age. To enumerate the special causes that have operated from one generation to another would be an impossible task, in view of the extraordinarily superficial character of what passes for history among us. It is possible, at present, only to indicate the nature of the process through which the styles—in righteousness and clothes—come into being, run their brief course and pass away; and that is what the present book attempts to do.

In a considerable degree even this modest effort must be negative. The first step in an understanding of the genesis of styles must be the realization that no absolute standard of esthetic excellence anywhere exists; and the first step in an understanding of morality also must be the recognition of the same complete non-existence of any standard of righteousness that is not wholly a matter of tradition, custom, transitory vogue. Perhaps this will seem unnecessary in an age so emancipated from superstition as the present. But emancipation itself is very relative. All peoples are emancipated from some superstitions, but none has ever been purged of all. It is a question of degree. To some readers, much of the argument will seem obvious. They are accustomed, perhaps, to view morality as a herd phenomenon. "Mores" is a word with which they have been long familiar. It slips off the tongue with the facility of frequent use. But there are few who do not bring themselves up short, sooner or later, with an inevitable and saving "None-the-less." Up to a certain point it is easy enough to be advanced, to speak of righteousness as customary, to scout the necessity of religious sanction—provided only there be a point to draw the line. Such a line exists for almost every one, a line beyond which sophistication stops and dogma holds its sway. That is what justifies our argument. We are going to propose—with boring insistency, perhaps—not that righteousness is more or less customary, but that it is wholly customary; not that absolute standards of morality are hard to seek, but that no such thing anywhere exists.

So much for the causal aspect of the case. The effect of righteousness is equally important and vastly less familiar. Indeed, for many people, this is precisely where the line will fall between what is obvious and what is "preposterous"—as they may think—if not "intolerable."

PRE-VIEW

This is the side upon which Mr. Veblen showed that the net effect of any style is conspicuous waste and invidious display. The temptation is very strong for any one who is making a study of the social effects of righteousness to represent it as a deliberate conspiracy of the strong against the weak, the rich against the poor, the patriot against the foreigner. Not long ago some one "discovered" that international law countenances war, that it assumes war as a fact, and is chiefly occupied with prescribing the conditions under which nations may plausibly make war and the expedients with which wars so made may be "humanely" waged. In this sense international law is but one phase of the universal righteousness. Morality is the code whereby men determine whom to snub and when, whom to punish and how, whom to oppress and why, whom to annihilate and where. It is, in actual fact, the technique of odious comparison. None can be "good" except as others beyond the pale are "bad."

In modern times, under the modern codes, two forms of odious comparison are paramount: invidious relations between nations, and invidious relations between classes. Both are defined and at every point conditioned not merely by special canons relating specifically to war and the class struggle but by the whole body of organized morality. To give a realistic account of morality without showing this to be the case would be quite impossible. The orthodox theories of morals are unrealistic simply and solely because of their reluctance to admit these facts. Consequently this aspect of morality looms rather large in the present argument. No special chapters have been devoted to it. Rather is it too explicitly, in every chapter. Every chapter present, at least by implication and often all is one item in the technique of invidious comparison.

All this, however, is to be taken in a purely descriptive sense. From time to time the author's special sympathies may show. Who is quite without taint of moral parochialism? But in the main it is not the purpose of the argument either to approve or to deplore morality, or invidious comparison, or war, or the class struggle. To the eye of the student all these things appear inevitable. Man is what he is, and if one doesn't wholly accept the cosmos, "Egad, he'd better!" Nevertheless, inevitable as goodness is for civilization and the human race, there is some indication that it is not unshakably dominant over every single individual. The ascendency is general but not absolute. In the nature of the case some scattered few must be exempt—whether for their good or not, whether for the general good or not, we can not say. Changes do occur; and where they occur scattered individuals are cast up: moral strays; men without a country; the citizens of the world; the déclassés. They are the material from whom the heroes and the martyrs of our history have been drawn, and their strange, partially emancipated fate at least serves to throw into relief that righteousness in which the majority of the human race is at all times perforce content to grope.

CHAPTER II

MAN THE ANIMAL

BEFORE evolution came on the scene morality was usually attributed to some special seventh sense. The simpler and stronger moral reactions seem as vivid and immediate as nervous reflexes like the shock of pain. We are, simply, shocked. Just so, as "Moses" has recorded, no sooner had they eaten of the fruit of moral sense than Adam and Eve were shocked to see that they were naked. They were shocked exactly as we are when in our dreams we suddenly discover ourselves walking down Fifth Avenue in pajamas. "Why am I oppressed by sensations of disgust?" inquired the exquisite moralist of the eighteenth century, Lord Shaftesbury. "Because I have a nose!" Righteousness is nothing more nor less than a nice smell!

This assumption of a special sense of righteousness, almost coordinate with the physical senses, which, we must remember, were then less clearly understood than now, greatly assisted the evolutionists in constructing their natural history of virtue. When the evolutionary theory of the creation of mankind began to be discussed, in the latter part of the nineteenth century, the genesis of morality also became a burning issue. Man having become an animal species with wholly natural evolutionary origins, it became necessary to provide causes within the world of nature such as might give rise to each of his peculiarities, among them, of course, morality. Considering the setting in which this problem first arose as a subject for scientific investigation, no one need be surprised at the definitely evolutionary and even anatomical suggestions which at once appeared as the solution to the problem.

As an animal species man seemed to be possessed of many resemblances to his mammalian cousins, one being the mammary glands by which the young of these creatures are nourished. Some scientists have held that this method of nurture establishes a very tender relation between mother and young, marked by feelings of dependence on one side and of sentimental and protective attachment on the other. Out of this relation and the feelings it engenders grow all those complex acts of parental solicitude that have been lumped together under the so-called parental instinct; and this, in turn, has been taken as the basis of morality. For in a certain sense morality is altruistic. It inspires a certain regard for the interests of other people, and imposes a certain restraint and even at times positive discomfort in the interest of others. What could be more plausible than that an "instinct" which leads a mother animal to give up her life if necessary in defense of her litter is the evolutionary forebear of the moral sense of man whose greater love is expressed in laying down his life for his friend?

Viewed in the light of later criticism the difficulty with such a delightfully simple, though mysterious, theory of morals is that no such sense is anywhere perceptible in man. The physical senses no doubt exist. We are able by chemical reactions in the nasal membranes to differentiate Camembert from Liederkranz. Some substances by the direct action of the nervous system will cause almost instant spasms of regurgitation. This is a true reflex and entitled to be classified as a natural and inborn "sense"; but this is not the stuff of which morality is made. Things which really "make us sick" are few, and rarely met in the ordinary walks of life. Nor did the matters which caused Lord Shaftesbury's disgust bear any resemblance at all to these disturbers of the peace. No doubt suckling is accompanied by pleasurable feelings and a limp relaxation of all nervous tensions. No doubt also the fire and fury of rage are governed by the internal secretions. The physical aspect of drooling sentimentality may be observed in cows, and in cats all the external manifestations of righteous wrath. But true sentiment and virtue militant have other aspects than external posture of the body. What objects and incidents are capable of arousing such reactions in one animal or another is in no wise accounted for by the identification of the organs of secretion and the circuit of nervous control of bodily posture. Man may possess the same instincts as other animals and be capable of no more sentimental attitudes than they; but he is interested in other and more complicated things, and those things in their most complex aspects are the locus of morality. The instincts, moral or otherwise, do not suffice to clear the matter up.

Thus we are driven at last to the realization that the mystery of morals, like the mystery of life itself, is a problem not of potency but of complexity. The whole of human behavior is compounded of virtue and unrighteousness, and we are not likely to untangle them without unraveling whatever factors there are which enter into the organization of human behavior as a whole. These are many and various, of course, but not totally impregnable. If we give over the search for any one uniquely potent clue, assuming that open sesames exist only in works of fiction, and set ourselves to survey the field with unexcited eyes, we find that human behavior does not wholly resist analysis.

As an animal species man differs from his relatives not so much in the possession of special organs as in the general integration of the organs which he does possess. Each of these, including even the highest nervous centers, man shares with other creatures. The respect in which he is unique, paradoxically enough, is the degree of his generalization. Whereas the others are superlatively adapted to the various special features of their several habitats, man is adapted primarily to adaptation. He is capable of nearly anything.

Coupled with this generalized anatomy there goes a similarly flexible behavior. Man does in fact adapt to a tremendous variety of activities. This variety, moreover, is to be found not only in the extremes of difference between various men and groups of men but even between the various undertakings of each single man. The generalized capacities of man extend through the whole of life. Men are not just born into widely different habitats which elicit various responses and then at once and securely bind them fast; whatever their circumstances, men continue to expand the variety of their experience from year to year, expanding their habitat indefinitely as they go from one complicated activity to another still more complicated.

The methods of this continuous adaptation include all the organic flexibilities of human anatomy and in particular all the tortuosities of the human nervous system. It is not necessary to be an anatomist to see what the process must be in its larger outline. On one side we have the dexterity of hand and tongue and general bodily posture by which human experience can be expanded; on the other we have the mechanism of organization. It is through this that morality—which is to say organized behavior—is made possible; and its positive and negative poles are habit and intelligence.

Much heated discussion has raged about these words in recent years, and especially about intelligence. In popular discourse the word "intelligence" has always had a meaning not so much precise as invidious. Without bothering to state to ourselves exactly what we mean by it, we have assumed that it is the peculiar attribute of man, and that the degree in which each of us possesses it is the true measure of his excellence and value as a neighbor and a citizen. Such being the case, we have become most anxious to obtain from science a scale upon which intelligence may be correctly measured, so that we may not only know our own measure certainly but may exhibit it proudly to our enemies. So the "intelligence quotient" has come into being.

But the famous I. Q. has not passed without

a certain flutter of disagreement from scientists who have pointed out that we are measuring something which we have been able neither to isolate nor to define. For one thing, to distinguish sheer intellectual capacity from various types of acquired skill (which of course must involve habit) is practically impossible. At all events it has never been done with anything like complete success. For another, there seems to be no such thing as "general intelligence"; on the contrary, intelligence itself is of different types. A man who does not adapt readily to visual situations may do extraordinarily well with auditory situations; while a man who is not particularly adept in sights and sounds may display an unusual degree of muscular coordination and adaptive skill. Viewed in the light of the facts, "general intelligence" is a myth. It means "general superiority"—whatever that may be.

Perhaps, indeed, the reason why intelligence and habit have been so perplexing is that they are not distinct organs or faculties at all. They are degrees of each other. We have in man a being capable of acquiring habits. Now the acquisition of habits is the process of fixing upon the organism a mode of activity not ready made in the hereditary structure of this creature. Thus the process of habit-fixation itself presumes a capacity for novelty. Indeed, the novelty—man's ability to meet new situations with hitherto untried activities—is an indispensable part of the process by which those new activities are repeated, stereotyped, and so committed into habits. If we pay attention to the series of repetitions by which a novelty becomes a commonplace, we see that both novelty and commonplace belong to the same series. The difference between them is only between the early and later stages of the same process. Intelligence is the primary phase of habit; habit is the consummation of intelligence.

Thus we view ourselves as beings organically capable of a range of behavior that is uniquely broad but is continually limited and at the same time organized by our unique retentiveness. These two capacities depend upon and check each other. Viewed without illusions, man is a noble but baffled creature. With his retentive memory and order-loving habits, he would achieve a miracle of stability and secure control—if it were not for his fatally restless temperament, his incurable itch for innovation. Similarly, with his marvelous adroitness and ready flexibility, he would win the freedom and independent individuality of gods—if it were not his infirmity and curse always to go about repeating to the end of time not only his most felicitous adventures but his most unhappy strokes as well. Man is addicted to society and also to revolution.

Such being the nature of our behavior, such is the nature of our morality as well. We may conceive morality as order, achieved and permanent. In that case we have to reckon with all our sins of individuality. Or we may conceive it to be the freeing of the soul from bondage; and in that case we must be continually appalled at the ease with which we forge ever anew the chains of custom and convention. How we love our freedom! And how we love our chains! It is small wonder that even Jehovah regarded us with indecision.

CHAPTER III

THE BASIS OF DECORUM

MAN is a creature of habit and intelligence; but righteousness did not arise from either habit or intelligence. No more did any other aspect of civilization. Only a creature similar to man in his organic make-up could conduct a civilization; but not even man could invent one out of hand. Civilization is a system. So is that phase of civilization called morality. The basis of civilization is civilization, and the basis of righteousness is decorum.

On the morning of the eighth day, as we may presume from the ancient Hebrew epic, Eve having not as yet been thought of, "The Lord God commanded the man, saying, Of every tree of the garden thou mayest freely eat; But of the tree of knowledge of good and evil, thou shalt not eat of it." It would be interesting to know in what language this conversation was conducted, in as much as the possession of a lan-

THE BASIS OF DECORUM

guage itself requires not only knowledge but knowledge of good and evil. A language is a system of decorum.

According to one theory God created all living things in the fullness of their powers. This ingenious notion, which has been much quoted recently, was devised by a British clergyman of the nineteenth century in order to justify to himself his love of botany and his uncomfortable knowledge of the facts of evolution. He knew, for instance, that an adult tree trunk consists of a series of concentric rings each representing a year's growth. Thus the question presented itself to his mind: Were the trees of the Garden of Eden devoid of rings, or were they like the trees we know? Assume that they were perfect and complete trees as we know trees, complete with growth rings, and you have a scheme for reconciling Genesis and evolution. You have only to postulate that everything else was just the same. The trees were only just created; but they were created perfect, as though they had grown from seeds. The earth was created as we know it, in geologic strata complete with fossils, as though it had been born of a solar eruption and formed by planetesimals. In His divine omniscience God foresaw the theory of evolution and was thoughtful enough to provide all the evidence, embedded in the product of His six days' labor. Thus the first man was brought forth not only with a full complement of bodily organs complete with navel, though he had no mother; he was also miraculously endowed with a system of behavior. Ever since this time men have had to learn how to behave in childhood. He alone was created behaving, in full flight, speaking the Hebrew language (if that was it) as one to the manner born.

If any one is disposed to inquire what was the origin of righteousness, we are justified in parrying his inquiry with a question—What was the origin of language? Grant language, and you have granted a system of behavior. Grant a system of behavior and you grant righteousness. Every schoolboy knows to his perpetual sorrow that language is almost exclusively a matter of right and wrong, right pronouns and wrong verbs. Language is a perfect model in miniature of morality. It is from first to last a system of propriety. A solecism is only a breach of

THE BASIS OF DECORUM

etiquette, perhaps; but what about bad words? What about obscenity, blasphemy and prevarication? This is not a pettifogging question. Speech is not a mere means to occasional indiscretions: no conception of language is possible which does not give rise to them at once by an inward logical necessity. Language is a system of communication, and communications must be either true or false. A communication which is neither true nor false is not a communication; one which is, is righteous or wicked, according to the circumstances.

As a model of decorum language goes even farther than this. Of all the purposes communication serves, decorum easily outbulks all the rest taken together. We are thinking now not merely of the fact that the meaning of every word is established by usage, but of the further fact that the "dictionary meaning" of a word usually has little or no relation to its significance in actual communication. We say, "My dear sir," in order to imply the absence of any personal connection. We say, "Thank you so much," when we mean, "Please desist;" "You big bum," when we mean, "Charming fellow"!

These are set phrases, stereotypes, clichés, "rubber stamps." Most speech consists of set phrases. The extraordinary thing, however, is not the mere setness of the phrase but the peculiarity of its recital. "Our Father who art in heaven" is a set phrase, no different in kind from any other; and its use is typical of all. Practically no one ever considers the dictionary meaning of its words or the logical significance of the group of words. In such a phrase people mean to strike an attitude, to convey a certain sense of whom and what they are. It is not the same sense for all people, of course. Pronounced in a professionally elocutionary manner, such a phrase goes to show that the speaker belongs to the profession and requires to be treated as such. Mumbled under the breath, half inarticulately, the same words indicate that the speaker is a respectable and upright person, more or less addicted to property, and laudably hostile to the soviet form of government and companionate marriage. Rattled off with an unarticulated rush of tumbling syllables they mark the speaker as a member of a religious establishment which lays more stress on the performance than on the

THE BASIS OF DECORUM

manner of performance; he is a communicant in good standing, but quite free of morbid preoccupation with spiritual matters, and may be a prize-fighter by profession. In each case the words serve only to decorate an attitude. They have no other meaning to the people who employ them so, and they are almost never employed in any other way.

So extensive is this form of speech-decorum that the eminent American archeologist, Sinclair Lewis, has been able to reconstruct extensive harangues—the speech of "The Man Who Knew Coolidge"—in which nothing is ever said except by way of incantation. But the discovery is not a new one. Many authors have presented us with characters who go through life saying only "what seems wise in consideration of the effect on other people"; and history is full of eminent personages—Chesterfields as well as Machiavellis—who have made "guarded speech" their rule of life. Thanks to advertising and political reporting, the public has lately learned to identify such manifestations when carried to senatorial excess. This is the field of ballyhoo and buncombe. The origin of the good word "bunk" is especially informing. "When, in the sixteenth Congress," says one authority, "the 'Missouri question' was being discussed, Felix Walker, a member from Buncombe County, North Carolina, persisted in speaking when the house was impatient to vote, he was implored to desist but would not, declaring that he must make a speech for Buncombe." To this we need only add that Buncombe is the largest constituency in the world.

Remarks addressed to buncombe are apt to be specious, of course. But most human discourse is specious: precisely that is the burden of this argument. Moreover we must resist the temptation to suppose that buncombe is anything more than specious. In some cases the temptation is very great. An American president charged with diplomatic negotiations with our war-time allies states in public that he knows nothing of the "secret" treaties by which for some years those allies have been chiefly guided. A senator from Ohio states that the Honorable Harry Dougherty is "as clean as a hound's tooth." The attorney for Mr. Harry Sinclair in his two trials for conspiracy against the government states in the first trial that his client had no knowledge of the Continental Trading Company, received no money from its operations, and paid no money to ex-Secretary Fall; and in the second trial offers the defense that the money paid by his client to ex-Secretary Fall out of the bogus profits of his dummy corporation, the Continental, was in purchase of a third interest in a ranch worth notably less *in toto* than the sums paid. Our temptation is, in such instances, to say that the men are liars. But the case is not so simple. These men are not liars. Not one of them would have misrepresented his golf score.

The whole matter turns upon our understanding of what is the purpose and function of human speech. Its chief purpose is to serve as an expression of decorum. A soft answer, so we have been told, turneth away wrath; and the implication is that the decorous thing to do is to give such an answer as will serve to placate, not an answer which stickles over the logical niceties of what is technically true. The state of mind of presidents and senators, attorneys, pedagogues, parents, children, employers, servants, clergymen and contrite sinners is only the natural and familiar one of wanting to produce salubrious results. The sinner wants absolution. The clergyman wants conviction of sin. The servant wants to avoid blame; and the employer wants to evoke loyalty. Parents want to deceive their children into virtue, children want to trick their parents into kindness. An attorney wants to get his client off; a senator wants to support his party boss; and a president wants to be regarded as a noble statesman, not as an intriguing politician.

Parents are more numerous than presidents, and consequently they are apt to think that their little subterfuges of decorum are sincere or at least natural and unconscious, while the prevarication of a president is deliberate and sinister. But they have never been presidents. It is just as natural and commonplace for an attorney to hoodwink a jury or for a president to mislead his senate as it is for a parent to spin fairy stories about the stork. They are no more conscious of wrong-doing. Indeed, it is very doubtful if they are actually guilty of anything more than habitual, lifelong, bred-in-the-bone decorum. Buncombe is natural to them only because it is natural to the entire human race, and it is sinister in them only if it is sinister in the entire human race.

Listen to the casual conversation of a pair of clerks at lunch, or a group of housewives conversing over their teacups. Is their speech notable for its stalwart adherence to the truth?—its rigorous critical scrutiny of meaning?—its determined rejection of vagueness and innuendo and sloshy sentiment? Or do they give the impression that every word they utter is uttered because it is "appropriate"?—appropriate to their age and sex and station, to their respectability, to their reputation for wifely sentiment or for business acumen? Not only that. Disregard for a moment the actual burden of their present words. Do they give the impression of an earnest effort to apply critical intelligence to everything they say and do, or do they appear actually to strive to be like other people of their class and status? Do they seem to yearn to say what is expected of them—or the reverse?

The fact is that the interest in precise truth is sharply limited. As a general interest, extending through the whole of a man's affairs, it is practically non-existent. But, on the other hand, it is never wholly absent. It derives from inanimate things—from materials and tools—in contrast to decorum which derives from institutions and personal relations. Civilization is bounded on one side by machinery and on the other by decorum; and human behavior oscillates between the two poles, positive and negative. What we mean by truth—the kind of truth that is no respecter of persons and will be uttered though the heavens fall—is instrumental truth. It is derived from our use of tools and instruments. It is tested by the standard of "consistency" which is nothing more or less than a principle of mechanics. If things fit together and work smoothly without friction, they are consistent, true, mechanically efficient. The object of such truth is to get things done.

Originally and primarily the object of instrumental truth is to get mechanical chores done. Every one has such things to do; and therefore every one is interested—to that extent—in truth. The housewife whose conversation is one long uninterrupted stream of buncombe will be none the less rigidly meticulous about the number of eggs to put in a cake. She needs no instruction in logic or scientific method to induce her to test recipes experimentally. Even here decorum is never completely absent, naturally. The experimental attitude—that is, intelligence —is always suffused with habit. What cake is appropriate under the circumstances, the standard of richness and lightness and shortness, the dietary contrast it is designed to give: all considerations such as these affect the making of the cake and impose upon it the dictates of decorum. There will be even a set ritual for the compounding of the ingredients most of which will be sheer decorum: a technique that has been accepted on tradition, never tested and never questioned. Still, the eggs are scrutinized before they are added to the dough.

Nor, on the other hand, is decorum ever quite free from critical experiment, any more than habit is ever unmixed with intelligence. However staunch an adherent of all that is appropriate, the housewife may scheme to marry off her daughter with a clear-headed conception of the relation of means to ends worthy of a master mechanic and an adroitness worthy of a politician. The attorney does not recite his fictions in a vacuum; he studies his jury like an entomologist his insects, and tempers his prevarication to the fleeced lamb. The senator always talks like a senator; but the object of his talk is a party, and personal, victory.

The central point, then, is not the distinction of habit from intelligence, nor the contrast between truth and speciousness, but the nature of human relations in contrast to machinery. Human relations are established by presumption. Material things submit to mechanical efficiency. Each can be courted only on its own terms. Any given social success—a marriage, or a political appointment—is in a certain sense a material fact and can be brought to pass only by the application of instrumental logic. But although this may be true of the adaptation of means to end, the end is after all defined only by presumption; after it has been achieved, the good of it is only in the esteem of other people, and the means which can be bent to the achievement of such an end are similarly ritual.

Inevitably so: that is what human civilization

THE BASIS OF DECORUM

is. On the analogy of bees we might conceivably picture a human society founded solely on material fact. The queen is designated not by pope or parliament but by the organic fact of her physique. The workers are physically sexless females. The drones are males. Social relations between physically similar workers are undifferentiated by any putative social distinction. Such a society is perhaps barely thinkable for man. But that is not the basis of human society. Human society is founded upon personal relations established not by organic difference but by supposition. In consequence of this basic fact, all the material of human intercourse must be of the nature of supposition, too. The lesson of civilization—the system which each individual must learn at the expense of his capacity for habit—is the lesson of supposition and decorum. If he learns the English language, then he discovers himself to be an "Anglo-Saxon." If he has an "American accent," then he finds himself an American. If he blurs his vowels and says "bean" for "been" (without saying "wear" for "were"), he is a true denizen of Park Avenue. If he says "toid" for "third" and "ahn" for "on," he is a native of the Bowery. Having swallowed these monstrous discriminations, any further casuistry of behavior or speciousness of language into which he may be led thereafter will be mere peccadillo, the mere expression of an inevitable decorum, perfectly natural and unavoidable under the circumstances. He has already swallowed the system from which everything appropriate to him must shortly follow.

CHAPTER IV

O TEMPORA! O MORES!

Two views of civilization are possible. One would regard it as a noble edifice founded upon eternal verity and raised to glittering pinnacles of lofty principle. The other would regard it as a sort of kitchen-midden or, to be plain, a common dump. One of these views is unflattering and is founded upon fact. The other is edifying and is derived from specious sentiment. The fact that such sentiments are the sentiments of cultivated people is a commentary not on civilization but on cultivated people. What is cultivation? It is a process of indoctrination by which certain standardized beliefs are substituted for the plain evidence of the senses. Some of these beliefs deal with esoteric matters like art, literature, music and gastronomy; others are still more esoteric and deal with the creation of the earth, the will of God and the whole duty of man. In each case the fundamental belief is that this belief, far from being a late product of the accumulating process, is actually the rock of ages on which everything else has been laid down. "And God spake these words, saying . . . Thou shalt not kill. Thou shalt not commit adultery. Thou shalt not steal. Thou shalt not bear false witness against thy neighbor." On the most conservative estimate this was several thousand years ago: yet "No Christian man whatsoever," that is, no communicating Protestant Episcopalian man, "is free from obedience of the Commandments which are called moral." So regarded, righteousness appears to wear a sturdy and uncompromising mien.

Fads, to be sure, do change with proverbial fickleness. But what are fads? They regulate only such things as the clothing of the body. Therefore, we think, they are only the clothing of society. We brush them aside, as unworthy of consideration. But in doing so we make a great mistake. Perhaps we more than half intend to deceive ourselves about them. We pay little attention to "mere conventions" because they fluctuate so rapidly; and we take small notice of the fluctuations because only conventions are

O TEMPORA! O MORES!

affected. In other words, we refuse to believe that anything important in civilization can change greatly; and therefore when we see anything changing we declare it of no importance.

Nevertheless, as philosophers have said, we are the slaves of fashion. Fads hold the world in their despotic sway precisely as the books of etiquette announce; in fact, more precisely. Your book of etiquette knows only the world of polite society, whereas convention rests equally on the polite and the impolite. We do not ordinarily think of the necessity which obliges a racing tout to wear a loud, checkered suit and compels a tough to spit before answering a question as being conventionality. The fashion magazines do not carry a page of information as to what the well-dressed gambler will wear this season, or a column on the proper demeanor of hijackers.

Consequently, we assume a sharp distinction between what is proper, and what is perhaps appropriate to the lower classes and the demimonde, overlooking the fact that our notion of what is proper is after all nothing more than what we conceive to be appropriate to the middle (and upper) classes, and that it is certainly looked upon by those who wear blue flannel shirts and eat peas with their knives as made up of nothing in the world but gross affectation. But if conventions are the affectation of one social class, it seems to follow that the affectations of other classes are conventions, and presumably conventions of exactly the same sort, to wit, rules of behavior regulating all the lesser affairs of life.

Among students such conventions are now called folkways, out of deference to William Graham Sumner, whose book, *Folkways,* has supplied the students of the social sciences with at least two words which are already in universal use and seem likely to remain a permanent addition to the vocabulary of the American language. Folkways is one of them; the other is mores, the Latin word for customs, most frequently quoted in Cicero's exclamation, *"O tempora! O mores!"* (The Latin singular of the word, *mos,* is such a shock to English-accustomed ears that it is seldom used.) Sumner's theory was that the folkways and the mores together rule society. They alone determine what may or may not be done by high and low alike. The folkways cover

O TEMPORA! O MORES!

the lesser affairs of life, while the mores form the back-bone of the most solemn duties and prohibitions of morality. Both are wholly customary.

To many of us this idea itself comes with something of a shock. We are quite used to regarding conventions as customary, if we ever distinguish customs and conventions at all. That is why we regard conventions lightly—because they are "only customs." The theory of *Folkways* seems to be that morals as well as etiquette are only customs. But there is another way in which this theory can be taken. We assume that conventions are trivial because they are customary; and therefore we assume further that if morals are customary too, they are also trivial. But if we begin at the other end, assume that morals are of the greatest possible importance, and that they are customary, and then move up to the proposition that folkways are customary too, there is no reason at all for our concluding that the folkways are trivial, because they are customary. On the contrary, reasoning in that direction we should properly make the inference that, folkways being customary in the same sense as mores, they are just as important if not just as sacred as the mores. Indeed, Sumner's distinction between the two is made entirely on the ground of popular prejudice in the matter. For Sumner, folkways and mores are exactly alike in kind. The only difference between them is that people think some customs important and others unimportant.

The reason we think morals are more important than manners is because we think they are also more basic, or more central. They are, supposedly, the fundamentals of our civilization. If we were challenged, we should probably reason the matter out in this wise. Civilization is an organically related system of institutions. These institutions are: the family (which orators place at the head of the list), the church, the state, the town, borough or other primary and local group, and so on. All these institutions are bodies of persons organized for carrying on some one or another of the major concerns of civilization which we classify accordingly as government, religion, the procreation and rearing of children, et cetera. Such major concerns, and such major institutions for their conduct, are defined by a compact body of major precepts and formulas: monogamy, loyalty to government, reverence to God, and the rest. These precepts are the basis of morality and civilization. Morality and civilization rest upon them, so we say, and do not rest upon the trivial customs relating only to outward behavior. Therefore we refuse to believe that the precepts of morality have no more substance than "mere custom."

But this argument, valuable though it is for impressing children, overlooks something which students can not afford to overlook. All these institutions and the concerns with which they deal are artificialities. Church and state and family are not separate growths of different species, like raspberries, and gooseberries, and strawberries. They are generalizations, distinguished from one another, or indeed separated at all, only by our thinking. Some one invented these distinctions, long enough ago so that it has become customary for us to make them. Not only are they artificial: in certain conditions they might even be fictitious. Church and state are not everywhere two distinguishable institutions.

They were not, for example, in ancient Rome. They were not only recently in Mohammedan countries like Turkey. Family and state are quite indistinguishable among most primitive peoples, such for example as the Hebrews of the time of Genesis. And even among ourselves these divisions are quite arbitrary and even accidental. The awe in which we hold the national flag and the person of the chief executive is religious in character. Catholics remove their hats when passing their church; patriots remove theirs when the flag goes by. Every aspect of family life is now regulated by the "state," so that whereas government was once a function of family organization, now family regulation is one of the functions of government. In short, all these complicated relationships intermesh; they are all part of the same machine; and where the engine stops and the transmission begins is a matter that is determined entirely by the convenience of the mechanics. To the architect the family is the institution which functions in the residence; the state that which holds forth in the state-house; the church that which occupies cathedrals. If we add the Masonic Temple, we

O TEMPORA! O MORES!

have another institution that is part family (since it consists of "brothers"), part church (since it conducts solemn and exclusive ceremonies), and part state (since it taxes its members, swears them to fealty, regulates their manners and morals and assumes responsibility for their wellbeing). In other words, our institutional system has been overlooked by the Bureau of Standards. If morality is derived from these institutions, then morality is a confused and conglomerate business.

This confused state of affairs would be incredible if we were to assume definitely and finally that civilization is organized from the center outward. On the other hand, if we make the opposite assumption, then the failure of our institutions to live up to the preconceived notion of cleanly chiseled foundation-stones is not only explicable but natural. This assumption is that instead of the folkways being derived from mores and mores from the basic institutions, the folkways are really the origin and source of everything. The folkways come first. The mores are the customs which people believe to be most important. They form morality. The various orders of association into which people enter are those which arise from the commonest and "merest" customs. They form the institutions, basic and otherwise.

We have only to reflect that more men are kept straight by good breeding than by prohibitory laws to see how much can be said in support of this apparently topsyturvy theory. The notion that morality arises from the deep foundations of society gives us a picture of people going about the world refraining from theft because it is contrary to the fundamental laws of society, refraining from murder because it is contrary to the fundamental laws of society, refraining from adultery because it is contrary to the fundamental laws of society. But as a matter of humdrum fact, we know that this is not the case. Very seldom in any man's life does he have occasion to give a thought to the fundamental laws of society, whatever he might take them to be when the rare moment came. People refrain from little thefts because it is not decent, not well bred, to be seen picking up other people's handkerchiefs; and such ideas of decency become so ingrained that no one thinks of doing underhand things even when others are not looking.

The principal reason so few men steal a million dollars is that these same folkways see to it that almost no one ever runs across a million dollars lying on the curbstone. Almost no one commits a murder because the folkways make it ill-mannered to lose one's temper to the point of saying "Damn!" The folkways also deprive us of deadly weapons. Where every one goes about armed to the teeth murders do occur. Where it is both ridiculous and mildly immoral to carry arms, very few people ever shoot. The chief safeguard against sexual immorality—let us be perfectly honest with ourselves!—is the code of etiquette which forbids a gentleman and a lady ever to be alone together under circumstances even in the slightest degree favorable to temptation. This is what we have vaguely in mind when we deplore the unconventionality of youth. We admit readily enough that the young people's present intentions are innocent; but we do not propose to run any risks. This is how society is governed by the folkways.

If Sumner is right at all, every civilization has

consisted of folkways. That means that all history is the continuous, perhaps evolutionary life of folkways. The folkways strain has been in existence as long as the human strain. At any given point we make a cross-section of both, and what we find in the cross-section is a complete organism (man) and a complete system of folkways (civilization); and each is descended from earlier stages that could be seen in earlier cross-sections. In this Sumner is borne out by all the evidence turned up by the science of anthropology, the study of the continuous development of human civilization, of which, indeed, Sumner's book is only a sort of literary embroidery. Assuming this continuity, we must recognize that the sanctification and reverence of morality are already in full swing whenever and wherever the mores are in full swing. That is what mores are: the folkways which determine what people shall reverence. There are some customs which deal with casual and trivial affairs. This means, being translated, that there are certain folkways prescribing how ladies shall sit; and there are other folkways prescribing that these folkways shall be regarded as customs. One

O TEMPORA! O MORES!

folkway tells little boys not to cry, as it is not manly to cry; another folkway directs mothers to chide but not to beat little boys for crying, as it may be unmanly but not deeply sinful to infringe that folkway. On the contrary, another folkway forbids a colored man to make protestations of love to a white lady; and it is coupled with a further folkway which directs the male relatives of the lady to lynch the negro who breaks this moral law. If this reasoning holds, what folkways are important—and how important—is determined exactly as it is determined whether one should spoon soup in or out. Each folkway arrives, delivered to us in childhood by our mentors, surrounded by other folkways which show us how to take it. Mores are surrounded with awful warnings. That is the package in which they are presented.

Whose signature is on the package? Who guarantees the genuineness of folkways? The folkways themselves provide the answer to the riddle of their origin. "God" frowns upon miscegenation—so we have been instructed from time immemorial to believe. "The Pope" discountenances scanty clothes. "Grandma" doesn't like little boys who cry. Some of these agents of righteousness—the bogy-man who plagues naughty children who get out of bed at night; Santa Claus who rewards good little children with presents once a year—are rather thin. Adults have a saying that there isn't any Santa Claus. But ask any one of these adults what brought about Prohibition. If he opposes it, he will tell you it was forced on a reluctant public by the Anti-Saloon League and the Rockefellers. If he favors it, the agents will be the Anti-Saloon League and God. In neither case is he at all likely to go behind these active agents to the gradual seepage of sentiments which in the end scarcely more than flowed through the obvious and personal channels in their down-hill run to the lowest level of common humanity. Yet those sentiments are historically undeniable.

What brought them about? Sumner says "adaptation." That is a professor's way of saying that he doesn't know. It means "they happened." Conditions change, and folkways change. Is there any universal law stating what conditions rule what folkways, and what direc-

O TEMPORA! O MORES!

tion adaptation has to follow? If there is, it has not been discovered. Perhaps each case is a law unto itself. In the case of Prohibition we may be able to notice certain signs. In the archives of one of the oldest eastern universities there lies a precious document, the contract between the builder of the first university building and his laborers, signed and executed while the state was still a loyal colony of his most Christian Majesty, George III. The contract stipulates that there shall be kept always on the premises so long as the work is going on, accessible to all the workmen, a barrel of prime quality Jamaica rum. Such were conditions in the American colonies. Stop anywhere along Fifth Avenue where a sky-scraper is going up and search the premises for rum. We shall not find any for at least two potent reasons quite outside the field of sentiment. One is that men who are soaked in rum are not safe men to have around the place. The modern sky-scraper is a dizzy place to work. Even the common laborers who merely trundle brick and mortar have to push their barrows across open-sided platforms hundreds of feet in the air, while the structural steel workers ride huge beams swung out over space from the thirty-seventh story. A rum barrel is not a safe companion for the engines that control these derricks. Moreover, the regiment of workers who raise such structures are a highly organized machine themselves. To perform effectively they must be every man jack in his place. If half a dozen firemen, fraternizing through the week days, staged a big party on the Sabbath, and consequently failed to turn up on Monday morning, the whole regiment would be effectively disorganized. To function properly, the hands of a modern industrial enterprise have got to be one hundred per cent. sober.

That is, the workmen have got to be sober. The owners may spend the winter in one continuous round of inebriety under the sheltering palms of Cuba. What is terribly wrong for one class may be right enough for another. In one sense this difference arises from a quite material fact: the actual physical differences of occupation. What impairment of the functions of an absentee owner is worked by even the most chronic alcoholism? That, perhaps, is the business of the absentee owner—to live like a gentle-

O TEMPORA! O MORES!

man—to be as drunk as a lord. But a laborer, under modern machine conditions, has got to stay sober, out of respect not merely to his masters but to the safety of his own skin. Even owners discountenance driving a car when drunk; and tossing red-hot rivets is an occupation quite as hazardous.

But owners also discountenance too much conviviality on the part of working men at any time. This is a difference of another kind. It is the business of a working man, so owners reason, not to waste his meager substance in riotous living, but to live soberly, industriously, contentedly, on his twelve dollars a week, making the ends meet nicely and never listening to the seductions of "foreign agitators." The foreign agitators, too, and even the domestic labor unionists, are very likely indeed to make their headquarters in the "working men's club": the corner saloon. For some years now the rumor has been circulating that the chief motive of the millionaires who financed the final, successful drive of the Anti-Saloon League was a desire to close the corner saloon and that their chief interest in doing this was not so much to exclude alcohol from circulation as to eradicate the working men's club, to stamp out a center of radical and unionist agitation. So conceived, prohibition was a strike-breaking device. This may be true or it may not. The millionaires whose benefactions to the League are most often discussed are also known to be hostile to organized labor. But what millionaire is not hostile to organized labor? Perhaps the whole thing was just a coincidence. Still, even as a coincidence it is a most informing one.

Whatever their specific causes, mores always arise from "certain conditions." On one side these conditions are invariably physical, and indicate the behavior which is clearly necessary under the physical circumstances. But on another side the conditions are social and institutional, and indicate which classes are to be affected by the physical conditions—and how.

CHAPTER V

GOOD TASTE

LIKE falling rain, the mores descend upon the just and the unjust, the strong and the weak, the rich and the poor. Consequently it is easy to ignore the fact that the mores which descend upon different classes are different mores. That is the very essence of the mores, of course; but we can overlook their essence in favor of the manner of their falling, since in method of application all mores are alike.

So are standards of taste. In the method of their application, all types and varieties of taste are exactly alike. But whereas one can think of mores as the compelling force behind morality, without bothering to consider what the various mores are and how wide is their variety, one can hardly imagine taste, or the various goods that satisfy it, without thinking of actual, concrete appetites and actual, concrete goods. To be sure, life is easier to bear if we do not think about it at all, but only dream of all things working together for good—a vague, mythical abstraction, perhaps, but still something quite outside ourselves. Then life seems anchored to the bed-rock of a quality not of ourselves but of things. If we think of value only to congratulate ourselves that since everything which any one desires is possessed of this quality in some degree, we do seem to have a basis for concluding that though mores come and mores go, value and goodness remain for ever the object of our desire and the goal of our pilgrimage. But how, on such a basis, are we to discriminate among our whims?

From a sufficient distance not only do all tastes look alike: all men look alike. They all want the same things, namely life, liberty and the pursuit of happiness, to use the quaint phrase of an early American philosopher. As Max Eastman once remarked, every man desires to support himself and if possible his family. That is, every man is an animal; and as an animal he shares with all other living creatures certain "unalienable" attributes. He is alive. Furthermore, to enjoy any other unalienable right he must stay alive. To do so he must eat; he must sleep under cover; he must protect his body. As an animal he is even committed to the mammalian system of procreation—what Mr. Milt Gross calls "de stoic mettud": he therefore has near relatives and a stake in their metabolic solvency. "Fundamentally" all men do want the same things—if we look at them from a sufficient distance.

But the distance at which all men look alike is one hard to maintain. It brings man into unpleasant focus as a higher animal. If we fail to relish this view—and we do—there are two things we can do about it: we can extend the distance, or we can shorten it. By extending it, we blur the entire foreground, man, animals and earthly habitat all together, and at once enter the realm of philosophical abstraction. This is a land of Latin capitals. The good is now the Summum Bonum. Of course, in the nature of the case, the Summum Bonum is too far off to be clearly visualized. It is rather vaguely sensed as a direction, or as a limit in the mathematical sense of the end of an endless series. The Highest Good is the essence of which each little good, however trivial, in some degree partakes. The Highest Good is the unrealized object of every human wish. What is goodness? It is the attribute of every good thing. Whatever is desirable is to some extent good; and The Good is that quality in its purest and highest form.

All this is perhaps true. Furthermore, it has a pious tinkle and recalls to mind the name of Plato. Consequently, like the diamond bracelet of Lorelei Lee, it makes us feel very, very nice. But it nevertheless leaves one thing to be desired: it leaves us quite impotent to identify any single concrete, salubrious experience.

On this point at least the medieval scholars had the best of us. By identifying goodness as the will of God they saved it from the vagaries of human will. But they, too, escaped from the specific into the abstract. Soon they found themselves confronted with this conundrum: is goodness good because God wills it, or does God will it because it is good? Unfortunately both of these questions lead to the same answer; and the answer in each case is final and unpleasantly abrupt. We can say that the good is that which God wills; and we can say that God is He who wills the good. But if we ask, "Why," in either case, we can only answer, "Because He does."

Suppose that goodness and desirableness are identical. Whether it be God's will or man's, the desire points out its fulfillment as a good. Suppose that the good is that which is desired and desire is a craving for what will do us good: then where are we? We want what we want. Instead of extricating goodness, we have plunged desire, too, into the bog. Desire and good are two sides of the same thing, but what is it?

So stubborn is this tautology that one of our eminent contemporary thinkers, Professor G. E. Moore, of Cambridge, has proposed that we accept the good as "indefinable." But this is what Mr. C. K. Ogden calls "giving up the problem in the grand manner." By treating the desire-good situation merely as a "unique" type of relation between man and the object of his interest,—the value-relation, which we symbolize by V,—we can go on and employ V in equations of all sorts, indicating the "asymmetry" of V, the "limits" of V, the "function" of V, and all the other antics of pure variables, until we know fully as much about V as mathematics has divulged about our old friends x and y. Thus we discover what it means to call the good an indefinable value-relationship. It means a lot of very interesting mathematics. But about the actual goods of life, which it is the object of morality to secure, we know as little as ever.

There is something about these fantastic formulas which suggests that we really do not want to know. A truth which requires so much metaphysics and mathematics to hide it must be most unpleasant. It is unpleasant—to our moral vanity; too unpleasant to be talked about. As we say, there is no disputing tastes. What we mean by this is first, that tastes differ, and second, that nothing can be done about it. Moral vanity is so intense that it prefers not to recognize the fact of difference, so intense that the only satisfactory way of dealing with those who differ with us is to "exterminate the brutes," as Joseph Conrad's hero wrote of his Congo friends. Therefore in club rooms, where mutual extermi-

GOOD TASTE

nation is impracticable, we declare an armed truce: *de gustibus non disputandum est.*

If we now bring mankind once more into sharp focus in the foreground, we can perceive with startling vividness an unpleasant fact: that men do not all like the same things nor hold the same things good. They do, of course, take a unanimous satisfaction in "life, liberty and the pursuit of happiness." But men are never merely animals. If they were, this unanimity would be significant. Cattle, also, have a preference for life; and life for them means plenty of grass. But men are unanimous only about life. They do not all live on the same things by any means. Indeed, they arrange to stay alive by inconceivably different expedients.

We find it agreeable to imagine that the heathen do not really relish the fruits of wickedness. If they could only taste the joys of civilization, how much would they not prefer them! This makes us out their superiors—since we actually possess those joys. But the fact is the heathen do relish wickedness. Cannibals actually enjoy dining off human cutlets. They consider them superior to pork and veal. They do not feel miserably debased, so debased that they must eat the most loathsome messes. They gloat over the prospect of a missionary stew as we do over a frozen pudding—which they would hardly feed to a condemned criminal. Mr. Sedgwick has given us a most vivid picture of Cortez's Indian allies dancing on their battlements, waving in an ecstasy of glee great lumps of meat carved from the ribs and thighs of their fallen foes. No feeling of disgust did o'er their spirits fall when they looked upon the Aztecs chopped particularly small! On the contrary!

What does most excite the disgust of savages is our innocent and simple pleasures. We have a great taste for touching each other with the lips, on the lips and elsewhere. It is to be presumed that we do it because we like it. Now and then we hear some one spoken of in terms of highest praise because he has through many years saluted thus an aged mother; but our praise does not imply that this performance was onerous. We do not bestow this praise as for a difficult and tasteless task which was nevertheless resolutely faced. No, we begin by supposing that the act was always spontaneous and pleasur-

GOOD TASTE

able; and we praise rather the fine taste of the man who has thus found joy in what a Chinese writer once contemptuously called our "non-amatory osculation."

To understand how a Chinese would feel about such a performance we can consider our own sickening disgust at the perverted values of the ancients. It is a fact that among the ancient Greeks and Romans close physical intimacy between men and boys was very common and—if we can credit such a thing—very enjoyable. Merely to think of it sends waves of nausea over us. We draw the veil of silence over this aspect of the glory that was Greece and the grandeur that was Rome. Yet this, let us consider, is precisely what a decent and refined Chinese feels of the perverse tastes of our most filial pleasures.

We would like to believe that decent things are always most agreeable—that our tastes are formed upon the pattern of our scruples. This would suit our moral vanity. Yet as good a case if not a better one can be made for the precise opposite, to wit, that our scruples, the mores, take their pattern from the things we like. An eminent economist once said that among us

[71]

chastity is a class prerogative. More strictly, it is the cherished possession of the middle classes. The poor can not afford it. They can not afford the elaborate subterfuges of chaperonage and segregation by which alone it can be maintained. While, on the other hand, the rich have no need for it. Their position is based on a social standing so high that little or nothing is added to it by a mean and petty prudence. What profit is it to a prince to cherish chastity? Nothing that he does can ever diminish the refulgence of his exalted position; so that he is in a position to enjoy both universal acclaim and the pleasures of untrammeled intercourse. From these facts of position and the values accessible to that position it follows that the behavior of princes is usually "free," the behavior of the middle classes is circumspect, and the behavior of the fourth estate profligate in the extreme.

For every variation of taste and value there are always compensations. The question was once put to Mr. Bernard Shaw whether it is possible for an actress to be chaste, to which Mr. Shaw replied, "Why should she?" As he went on to point out, she is déclassée from the moment she becomes an actress. Decent persons close their doors in her face. The censors actually substituted the word "actress" for the word "street-walker" in the subtitle of a film as conveying the same meaning euphemistically. Under these circumstances, what is the value of chastity to an actress? Obviously nothing. The burgesses of the middle classes place a high value on impeccability in these respects. They may thereby lose something of the freedom enjoyed by dukes and actresses; but in compensation they enjoy the privileges of respectability described so succulently by St. Paul. "The wife hath not power of her own body, but the husband; and likewise also the husband hath not power of his own body, but the wife. Defraud ye not one the other." Actresses may forfeit the advantages of this neat system; but as Shaw indicated in *Love among the Artists,* they obtain in compensation a freedom to refrain from any intimacy that has ceased to be mutual and spontaneous. It was the peculiar pride of Shaw's heroine that she never bestowed so much as a casual kiss without really feeling it. That is something.

If we turn to concrete things—the physical objects by which we express our simple tastes—we find a situation still less gratifying to our vanity. Whereas the other animals eat the common natural foods to which their digestive apparatus is directly adjusted, we oscillate between puffed air and triply larded pastry. The succession of our meals is marked by a succession of liquid stimulants and the space between given over to inhaled narcotics. Whereas savages wear furs or sarongs according to their latitudes, we muffle ourselves from head to heel in all climates and at all seasons, affecting a touch of fur in summer and an extreme of brevity in chiffon silks in winter by way of flourish. Nature having given us horizontal feet, we devise contrivances to set them at an angle of forty-five degrees. Nature having made us round, we invent straps and girdles to flatten out our contours. We create artificial daylight and then sleep while the sun is up. We lock our houses against intruders and install an instrument through which we are open to intrusion on every pretext at all hours of the day and night. Such are the things which we hold good.

In all these things, according to Mr. Thorstein Veblen, we are subject to the continual promptings of pecuniary emulation. A hat, we hold, must be tall and shiny. Why? Because it is fragile and expensive so. Coats and trousers must never be shiny. Why? Because that is a condition to which all coats and trousers naturally aspire. Gloves must be white, dove-gray, faun, because gloves of all things dirty quickly, and these colors will best advertise how little useful work we do and how often we can change our gloves. Automobiles must be capable of doing eighty miles an hour, although thirty-five is the limit fixed by law. As the advertisers tell us in their skilled promptings of our jealousy, it is the reserve power that counts—that is, the reserve pecuniary power. In their car we shall be able to show to all how much more we can afford to have than can be used by any one. Thus, says Mr. Veblen, money sets the standard of our values, and the touchstone of our goods is waste, not taste.

These principles affect us differently at the different levels of the social scale. For some the highest degree of excellence and the embodiment of all that we hold good is a Rafael or a Van Dyke, bought for huge sums from bankrupt Europeans. For others it is an inherited Corot. For still others it is a steel engraving of Leonardo's *Last Supper;* or, a colored print of Rosa Bonheur's *Horse Fair;* or, a tinted photograph of three apples and a fish (*not* by Cézanne); a chromo—"God Bless Our Home"; a crucifix with the corpus done in incandescent paint. Some eat garlic; others eat potroast and thrice boiled cabbage; others round steak and onions; some sirloin; some fillet of sole; and some pheasant out of season. At certain levels we make it our proud boast that the babies get a fine funeral when they die; at others we aspire to keep our infants alive. It is all a question of standard of living. Some people no more expect to ward off death than to be king; yet others of us make a slight paleness or indisposition of the "liver" the occasion for a hurried trip half around the world.

Is the standard of living a moral phenomenon? It would not seem so. No righteousness is attached to wealth, or so we have been told. It is thought to be harder for a camel to pass through the eye of a needle than for a rich man to enter the kingdom of heaven or escape conviction for conspiracy against the government of the United States; and blessed are the poor. Yet the emoluments of poverty are not universally admired. As we say, each to his own taste, however his taste was acquired.

In general it was acquired by birth, and by the force of long habituation—never, certainly, by precept. There is no precept which directs us what to like. We like what we can; we like what we are; and what we are is determined, as the mores are determined, by larger forces moving obscurely in the background. We sometimes call these forces economic; but they are no more economic than moral. Their economic aspect is one by which they can be measured. But their moral aspects are deeper and farther reaching because they embrace all that we hold good and all that we hold right. The fact is—whether we speak as economists or moralists—by civilization all men are created different; and those differences cut to the very bottom of life, liberty and the pursuit of happiness. They divide our tastes so far that there is no disputing them. They set us into nations, communities and classes; by their tyranny they dictate what things we shall hold good.

CHAPTER VI

FINE FEELINGS

STRANGELY enough, of all the fine feelings and exquisite perceptions which righteousness produces in the breasts of the most sensitive—feelings which, like good taste, are sometimes even regarded as the very source and well-spring of morality—the most typical if not the highest and most precious is a sort of cold ferocity. The chief mark of a scrupulously moral man is his capacity for righteous indignation. The righteous are like the Lord God of the Hebrews; they never hold them guiltless who have blasphemed against them.

What is strange about these sentiments, of course, is the contrast between their sanctity and their conspicuously retributive character. We could understand them if we could set them down at once to mob hysteria—"the ties of common funk" as Rudyard Kipling eloquently called them. But in that case why should upright people be so proud of them? Or, to put it the other way, if we were to define the moral sentiments in advance as the finest and noblest of which man is capable, we might perhaps form an imaginary picture of them out of such elements as pity, sympathy and understanding; but in that case we must inevitably be shocked by the fierce brutality of the actual picture. What is fine or noble in the sentiment, "He that killeth any man shall surely be put to death"?

Very likely the anomaly would be resolved if we could see clearly that this retributive ferocity is the first fruit of primeval savagery and that it has been progressively softened and sublimated by the advances of civilization and the deepening of morality. But precisely the contrary seems to be the case. Mr. Bernard Shaw has remarked that notwithstanding moral preachment, to strike a child in anger is not so beastly as to strike him in cold blood and then add to primitive injury the sophisticated insult that one does it for his own good, in the manner of Jehovah dealing with His chosen people. The contrast between primitive and sophisticated retribution is of just this kind. In the primitive tribe, retribution was swift and hot; among us it is slow and intolerably dignified. Not only is "he that killeth a beast, he shall restore it" still the sole basis of all law of property; we have invented the additional refinement of sending a man to jail for life for stealing a few coppers on the theory that he is a hardened criminal. Not only is "he that killeth a man, he shall be put to death" still the sole basis of our criminal law: we have improved upon the court procedure of the ancients, passing our accused through a seven years' pageant of hearings and rehearings, only to electrocute them (with all the consolations of science) in the end.

Still, retributive or not, righteous indignation is not self-interested. Possibly this is the essence of its charm. Jehovah was not jealous on His own account; He was concerned only for the piety of His chosen people. The death of Sacco and Vanzetti may give us a certain satisfaction (they were anarchists anyway!); but it can no longer edify the late South Braintree pay-roll messenger. Our pleasure is wholly vicarious—in seeing justice done; and we have no real ground for supposing that sentiments were different when justice was inflicted not by a public functionary hired for the purpose but by the relatives of the deceased. In that happy day bereaved relatives automatically became public functionaries, and we may just as well suppose that their finest satisfaction was in the contemplation of an arduous public duty laudably performed as to suppose the same of Judge Webster Thayer.

This vicarious element in righteous wrath is so significant that Professor Westermarck made it the basis of his great work, *The Origin and Development of the Moral Ideas.* The theory is that moral indignation, as distinguished from ordinary anger, is altruistic. It is indignation on principle, for the benefit of the community. Consequently it can be called the fundamental moral sentiment, and when it is brought into play we have the clearest possible case of purely moral behavior. Other motives appeal to various personal interests—be good and you'll be happy, and so forth—but righteous indignation serves no self-regarding purpose.

The violence of these so altruistic sentiments has always been well known and can hardly be disputed. What is not so clear, however, is their basis and their source. Is the code of moral behavior—the mores—sanctified and sanctioned by the violence and purity of these sentiments; or are the sentiments first instilled and later, when occasion offers, called into play by the code which fosters them? Does the code sanction moral violence, or does moral violence sanction the code?

We may not know—or we may say we do not know—*whether* the mores are the source of moral sentiments, but we can hardly plead innocence of knowing *how* morality sanctions violence. Whenever, for instance, we wish to arouse men to the highest pitch of frenzy (for whatever purpose, say the protection of certain investments abroad) we appeal not to their self-interests, but to their moral sense. Some say that all wars are economic; but statesmen know better. With melancholy cynicism they point out the lamentable lassitude with which the masses treat foreign threats to the investments of their countrymen. Men simply can not be induced to go to war for other people's property. They absolutely require to be inspired by the ashes of their fathers and the temples of their gods. Consequently it is necessary, when a war is on the cards, to equip them with atrocities. The statesman must manage to make out that the enemy is a ghoul who chops off the hands of little children and gathers up the corpses of fallen heroes to make fertilizer of them.

This technique of moral indignation was never more completely illustrated than by the sack of the Benjamites, narrated in the concluding chapters of the Book of Judges. It seems reasonable to suppose, upon the basis of our knowledge of modern wars, that somebody obtained some advantage out of the sack of the Benjamites, and that his interests may well have been instrumental in bringing the affair to pass. The chronicle is obscure on the point in ancient, no less than in modern, days. But the technique by which he brought it off is obvious. In the nineteenth century he would have sent in a missionary, who would have been eaten to the scandal of all good men, thereby necessitating a punitive expedition, and the cession of a treaty-port and a first lien on all customs duties. But in Old Testament days it was done thus. A Levite passed among the Benjamites with a concubine in train. He appears to have been a loafer, and made even this journey in suspicious circumstances; but it so happened that his concubine was ill-used by certain gentry, Benjamites, to such an extent that she died. Whereupon the Levite carefully packed her home, and there "divided her, together with her bones, into twelve pieces, and sent her into all the coasts of Israel."

This well authenticated, tangible evidence can be taken as the primitive equivalent of the Official Record of the Royal Atrocities Commission. As was expected, it aroused all beholders to the highest pitch of moral indignation. "And it was so, that all who saw it said, There was no such deed done nor seen from the day that the children of Israel came up out of the land of Egypt unto this day: consider of it, take advice, and speak your minds." The sequel came off according to form. The allied tribes descended upon the Benjamites and after some tough fighting wiped them out, men, women and children, so completely that wives had to be provided for the few remaining veterans by the sack of a misguided community that had remained neutral. Thus was the wrath of the virtuous appeased.

Incidents of this kind serve to show how complete is the sanction of the code. But public and concerted action is not a requisite. Many of our moments of most intense white heat are private and even secret, in the nature of the case. In the ordinary course of business moral fervor is seldom aroused. Even the administration of justice and the punishment of crime become reduced to a routine; so that, although it is supported by the strong sanction of the whole community, an appeal to the whole community for an expression of that sanction seldom requires to be made. Wars are occasions when such appeals do seem necessary. But behind the special crisis and behind the routine of law there is a zone into which the law can seldom penetrate where order is maintained and righteousness upheld by the stern virtue of the individual. This is the zone of the "unwritten law."

Sexual crimes in particular lie within this zone. In the nature of the case an appeal to public sentiment and public force is often extremely distasteful where the injury is sexual. Whatever the outcome of justice promptly done, a stigma would remain upon the innocent victim—a stigma that might be more momentous in its ulterior effects than the original injury. Thus secrecy is of the essence; and thus in such matters considerable powers are by general consent delegated to the individual. Cases are common of homicides quietly condoned on the ground of defense against or reprisal for some sexual violence; and stories circulate persistently of secret tortures and inflictions of the punishment of Abelard done in retribution for still more secret injuries.

This kind of righteous violence is universally condoned and even praised. Indeed, the sanction goes farther and practically obliges a show of violence by those who have been injured in the zone of the unwritten law. A man who failed of the proper pitch of indignation and was weakly prudent when the code sanctions rashness and brutality is thought to be no man at all. What woman would love a man who would not strike hard blows in her defense? We all know the feeling, and our military men traded on it not long ago when testing conscientious objectors. They formulated this stock question to be put to any man who proposed to be squeamish about war: If your sister were attacked, would you use force in her defense? If one answers this question, "Yes," then no ground is left (in the military mind) for objecting to any armed violence; while if one answers it, "No," one has branded oneself a half-man, emasculate and puny, contemptible for ever to all right-minded men and sound-hearted women. (Incidentally, no one seems ever to have thought of answering: My sister does not live in the vicinity of an army camp.)

The sanction of morality is not limited to negative expressions, however. It extends just as strongly, in a positive direction, to our tastes with all their differences and even to our standard of living with its prodigious inequality. This is what lies behind the problem of the sumptuary law. We are all familiar with the difficulty of placing a legal limit upon the consumption of any article in common circulation. The liquor problem has made us familiar with it, if we never were before. Professor Cohen quotes Spinoza as writing, "Those vices which are commonly bred in a state of peace . . . can never be directly prevented." Tastes which are commonly bred are not vices. They have the sanction of morality. Alcohol is frequently compared with opium on one side and with tobacco on the other. But in classing alcohol with opium the adherents of Prohibition are quite wrong, since alcohol is in common circulation among us and opium is not; while in citing tobacco anti-Prohibitionists are quite right. Tobacco is perhaps less harmful and alcohol is less common, but both are in wide and unceasing use.

Our difficulty with liquor is due to this peculiar confusion. Although drunkenness was once tolerated among us, for various reasons it is so no longer. But the forces which have outlawed the effects of liquor did not begin with the substance itself, and they have not really touched it yet. Even where Prohibition has been long established, liquor as a commodity and a taste is not severely frowned upon—certainly not frowned upon as opium would be in all parts of the country. We taboo intoxication and still value the intoxicant.

The inequalities of the social scale receive the sanction of morality both directly and indirectly and apparently have always done so. Such a sanction was embodied in the Mosaic code: "Thou shalt not covet thy neighbor's house, thou shalt not covet thy neighbor's wife [a household accessory], nor his manservant, nor his maidservant, nor his ox, nor his ass, nor anything that is his." The motive behind this commandment is clearly the preservation of inequalities in the enjoyment of property. We are not in the habit of coveting possessions inferior to our own. If there were any doubt on this point, the catechism wholly removes it. "What dost thou learn chiefly by these commandments?" inquires the priest; and the well-coached votary replies, ". . . my duty towards my neighbor. . . . To order myself lowly and reverently to all my betters. . . . And to do my duty in that state of life unto which it shall please God to call me." This is sufficiently explicit.

Our feelings on this point, furthermore, are intense—as intense as any manifestation of righteous indignation. Nothing arouses us more violently than a transgression of class lines. A royal personage becomes notorious on five continents for vulgarity as a result of doing what is only expected of every actress—making public appearances for fees and publishing syndicated memoirs. A negro, or a "native," is incontinently lynched for making such sentimental protestations as would be considered flattering in another member of the superior race; and as Mr. E. M. Forster has so vividly set forth, the advances may be quite apocryphal. A suspicion of impropriety is enough. Imagined improprieties have been quite as fruitful of lynchings in America as actual injuries. In Imperial Germany a man of common clay who failed to make way speedily enough for an army officer ran some risk of a saber in his stomach. Which should we regard with greater horror: the sight of a prince in oily overalls, or of a working man in a silk topper?

Strangely enough our earnest approval is attached not only to our neighbor's present wives and chattels, but even to his anticipations for the future. We labor under the belief that whatever a man, and especially a woman and her children, have been accustomed to throughout their past, they have a right to expect no less of the future. Their manner of life is funded, and becomes a fixed charge upon the world.

Even our courts acknowledge this, and fix the amount of damages, separation allowances, alimony and the income assigned from trust funds to the nurture of minor heirs, very largely on the basis of what will be consistent with the petitioner's "legitimate expectations." When foreign dukes and marquises are banished from their estates by an uprising of the rabble, our hearts go out to them. If we are bankers, we make them generous loans without other security than the hope of their restoration at some future date; and if we have estates at Newport and Palm Beach, we take them in as honored guests and compete with one another in supplying all the perquisites of aristocratic birth and breeding. Meanwhile, nothing arouses our unanimous indignation more speedily and hotly than the demands of "foreign agitators" for our laboring classes. These people actually want bathtubs and second-hand automobiles. A little more and they will be parading on Fifth Avenue, or even forcing their way into Trinity Church!

Just to think of such things nearly brings on apoplexy. We know what is right and proper. It is what we have always been accustomed to.

CHAPTER VII

GOODNESS SANS SUPERSTITION

ONE of the chief superstitions of the present age is that this is an age without superstition, and its inevitable corollary is the belief that in such a day of rational emancipation men can be righteous without prejudice. There are only two difficulties in the way of this belief: righteousness is of the nature of prejudice, and superstition we have always with us. Religions change. People who talk of the waning of theology in modern times are justified to some extent. Hebrew theology has doubtless waned. Our morality is less dependent on the Jewish Scriptures than it used to be. But it does not lack for substitutes.

Consider, for example, the theology of modern nationalism and its extraordinary resemblance to the tribal jealousies of the Jehovah of the Hebrews. "The national honor moves in the realm

of magic, and touches the frontiers of religion"
—so wrote Thorstein Veblen under the inspiration of the recent war in his great *Inquiry into the Nature of Peace.* Of all "lesions to the national honor," he went on, the most important are not concrete damages to persons and property but purely symbolic grievances: "disrespect or contumelious speech touching the flag or the persons of national officials, particularly of such officials as have only a decorative use, or the costumes worn by such officials; or, again, by failure to observe the ritual prescribed for parading the national honor on stated occasions. When duly violated the national honor may duly be made whole again by similarly immaterial instrumentalities; as e. g., by recital of an appropriate formula of words, by formal consumption of a stated quantity of ammunition in the way of a salute, by 'dipping' an ensign, and the like—procedure which can of course have none but a magical efficacy." Strange procedure for an emancipated age.

All this is very true—not only of international proprieties but of all proprieties whatsoever. These sentences describe all honor, all righteousness and all morality. In every civilization morality exists in partnership with superstition, or as Mr. Veblen says, with magic and religion. In all morality, offenses against property and persons gain a symbolic importance, and shade off imperceptibly into more excruciating offenses committed directly against decorum, ritual and ceremony. Contempt of court is a more heinous offense than petty theft; lese-majesty is more heinous than tax-dodging. Treason is worse than murder; and in a stable society blasphemy and unbelief (like that of the "Man without a country") is the most frightful crime of all, punishable by slow and lingering tortures and a long-drawn death.

Immorality shades off directly into sin. From simple wrong-doing we are inevitably led to symbolic impropriety. We come into conflict with one another; but in doing so we come also into conflict with the whole force of organized society, and we give offense to the mysterious forces by which society is governed. Such sins require interpretation, perhaps. As Mr. Veblen says of international peccadillos, "The common man is unable, without advice, to see that any given hostile act embodies a sacrilegious infraction of the national honor. He will at any such juncture scarcely rise to the pitch of moral indignation necessary to float a war-like reprisal, until the expert keepers of the Code come in to expound and certify the nature of the transgression. But when once the lesion to the national honor has been ascertained, appraised, and duly exhibited by those persons whose place it is in the national economy to look after all that sort of thing, the common man will be found no wise behind hand about resenting the evil usage of which he so, by force of interpretation, has been a victim."

So it is with every sin. The more excruciating the offense, the more mysterious it is, the more imperatively it requires official interpretation. As the mystery about him deepens the common man can only protest earnestly his good intentions and hope for the best. "We have left undone those things which we ought to have done; and we have done those things which we ought not to have done; And there is no health in us. But thou, O Lord, have mercy upon us, miserable offenders."

At the midnight hour of mystery, this darkness becomes so intense that even the interpreters must move warily. The uttermost sin of which a member of our Christian civilization is capable is the sin against the Holy Ghost. But what this is, is extremely dubious. Of all the figments of Christian theogony, the Holy Ghost is the least obvious, and perhaps the most potent. We know that the highest ecclesiastics show a special partiality for the Holy Ghost and appear to consider its good offices the most precious and effective. Lytton Strachey has noted that Cardinal Manning and the papal secretary never exchanged diplomatic notes without invoking the aid of the Holy Ghost for their peculiar enterprises. Sin against this agency is therefore most cataclysmically heinous.

But, like the soul-withering conclusion of a horrid dream, it is almost unintelligible. Saint Augustine declared this point of doctrine one of the most difficult in Scripture. Saint Thomas, the Angelic Doctor, interpreted it to mean three things. The first is direct blasphemy: doubting the fact or the efficacy of the Holy Ghost. The second is final impenitence. This is derivative, since a contrite heart is one of the trade goods of the Holy Ghost and lack of contrition can come only through the customer's sales resistance to the Holy Ghost. The third is all sins opposed to the characteristic quality of the Holy Ghost, that is, the operations of grace and sanctification of souls. But this is extremely subtle, as well as angelic. May blasphemy in the first degree be a sin of omission? In that case, who shall say what man has been sufficiently explicit in his acknowledgment so as to be absolutely on the safe side of unpardonable sin? Clearly, none but a priest, an authorized agent of the Holy and Infallible Church. In the second and third degrees this sin is one of commission; in the third it may mysteriously penetrate almost any otherwise venial offense; and who shall say when simple unregeneracy has reached the pitch of inexpiable sin against the Holy Ghost? No one, again, except the priest. There is scarcely a hasty word or untoward act which may not conceivably become, by force of interpretation, an unpardonable sin.

Thus in all climates and in all ages have men lived under the lash of superstition. In every

GOODNESS SANS SUPERSTITION

civilization the final appeal and last recourse is to the Levites. Why this should be so is a great conundrum that still baffles the scholars and historians. To hard-headed practical men the answer seems to be obvious: it is because shrewd and unscrupulous men have taken advantage of our weakness and organized a system of bondage to keep us in their power. Such, for example, is Brooks Adams' opinion of Moses and the obscure origin of the Hebrew-Christian system. Moses was, in his opinion, simply a diabolically clever politician who exploited the superstitions of the people—with a success to which we still bear witness. Such, again, was Thomas Hobbes' opinion of the Holy Catholic Church of Christendom. "Why are we priest-ridden?" he inquired; "Because we are timid fools!" This has been the explanation of stalwart free-thinkers in all periods.

But this explanation does not go far enough. What are the weaknesses that priests and medicine men exploit? Apparently they exist in us naturally, a perpetual invitation to exploiters. Theology did not begin with Moses, even assuming that he did in fact inscribe the Pentateuch.

It is as old as civilization. Sir James Frazer became interested in the mystic rites of Artemis connected with the "Golden bough"; and his explorations of the sources of these rites took him through many years, through all countries, and through twelve volumes. Mystic rites, and the obscure taboos associated with them, encircle the world, and embrace all peoples. Freud has suggested that they result from a universal neurosis inflicted like the sin of Adam upon all mankind.

These may not be the manifestations of abnormality, if anything so universal can be said to be abnormal. Yet they nevertheless require a phychological explanation. Morality and religion, taboo and totem, folkways and folk-lore, everywhere go hand in hand. Anthropologists can identify characteristic folkways and characteristic superstitions; and they may trace each to previous similar incarnations among other peoples. But they seem to be quite at a loss to say which dominates the other: whether taboos give rise to totems, or totems to taboos—and why. But is it necessary that either should come first? Can it be denied that religion and morality are two phases of a single process?

GOODNESS SANS SUPERSTITION

We are now dealing with sentiments; and in this field William James made a profound suggestion which quite possibly applied to the emotional expressions of religion and morality. His idea is generally paraphrased as the theory that we suffer because we wince and sorrow because we weep. But his own more careful statement was that the suffering and the wincing exactly coincide. Suffering does not follow wincing: it is precisely our total awareness of all the direct bodily responses of pain. We feel a wound, and we feel also all the inward tensions and outward cringing which the wound sets up. What he most emphasized is that we do not feel pain first and wince afterward: we feel the pain fully only through our act of wincing. No tears; no sorrow.

If we would pay a little attention to William James, we might save ourselves much confusion in discussing religion and morality. From what does religion originally spring: From the feeling of awe and "dependence on superior powers"? Or from the act of worship, ritual and ceremony? Or from the image of God "in the heart"? Or from a theory of divinity in the mind, from belief in certain intelligible articles of faith?

All agree that religion is impossible without emotions: the feeling of awe, the sense of inferiority, the attitude of trust, and so on. But according to James, no feeling is possible except in connection with action of one kind or another. For example, worship. The act of worship and the feeling of awe, do they not bear the same relation to each other as sorrow and the shedding of tears? And if this can be, is it any less reasonable to suppose that the image of God, the sense of His presence, our ideas of His character and meaning are similarly related and part of one, single, indissoluble complex?

The act of worship is not possible without an accompanying emotion, let us say. Neither is it possible without an accompanying stimulation of the imagination and even of the intellect. We worship. We feel our inferiority and our dependence; and coincidently we have a vision of omnipotent majesty, with an accompanying feeling of awe; and coincidently with all this our minds also are stimulated, and we explain or rationalize the whole matter to ourselves in words, the words of our belief.

If this be taken as the character of religion,

its relation to morality is at once apparent. The two merge with each other at all points and, as they say in geometry, "form one and the same straight line." Morality, too, is a complex of feeling and action. In action, the basis of morality is decorum, habit elaborated into ritual. In feeling, its basis is the sense of complacency and ineffable satisfaction with which we regard the good deeds of the faithful servant, and the indignation with which we resent the sins of the unrighteous. Such feelings are possible only for those of us who are ourselves in the posture of decorum. Complacency is the feeling that accompanies the correct performance of a rite. We feel complacent—and morally indignant—only when we are scrupulously acting in good form. Our morality, too, has its imaginative and intellectual elements. We have a vision of our elders, perhaps even of our forefathers, smiling approval of our scrupulosity, or turning in their graves at the impieties of others; and simultaneously we explain the whole matter to ourselves, rationalize it and justify it, by an elaborate theory of public benefit.

Thus analyzed, the elements of religion and

morality are clearly perceived to be identical. We may sometimes refer God more particularly to religion, and the ancestors more directly to morality; but this assignment is quite artificial. God and the ancestors belong together. They are, in a sense, one. In their historic origins they merge into each other: our god is always the god of our fathers. God is as much a part of the imagery of morals as He is of worship. The fathers are no less a part of our sense of divinity than God Himself. Our theory of "commonweal" is an article of faith. Decorum is ritual. Ritual is an application of decorum. Satisfaction with propriety is a religious sentiment, no less than awe. We thank God that we are not as other men are, extortioners, unjust, adulterers, or even as the publicans—for whom we feel the aversion that leads on to righteous indignation. The love of God and the wrath of God are very close together.

The essential identification of morals with religion has long been recognized by philosophers and theologians. It has sometimes been made the basis of an interpretation of religion. This is what the philosopher Kant meant to do when he proposed that the idea of God is conceivable only by the "practical" reason. Whatever may be conceivable by pure intellect, the practical activity of morals necessarily gives rise to the idea of God. Just as our tastes in all their variety nevertheless imply perfection as their ultimate superlative, so our sense of righteousness implies a Perfect Being. As we sometimes hyperbolically, say: virtue would be meaningless if we could not conceive of God as the supreme embodiment of all virtues.

As a matter of logic, this may or may not be sound. Kant is most careful to point out that it is not a matter of logic; it is a matter of "practical" reason—perhaps what we would call psychology. Our "psychology" is such that we must make God the focus of decorum and morality, of ritual and ceremony, of our self-satisfaction and our intensest indignation. We are, apparently, made that way. As one theologian has put it, if God did not exist, it would be necessary to invent Him.

Harald Höffding, the great Danish thinker, once raised the question whether it would be possible for mankind to be moral, to keep his faith in the triumph of good in a naughty world, without symbolizing this triumph in the person of an imagined God. He wrote before the day of psychoanalysis, and therefore before the theory of our universal inferiority complex and our cosmic Œdipus complex had been propounded to explain our inveterate craving for divine assistance in the works of man. But his answer was negative none the less. Our craving for images and symbols is too strong—as strong, apparently, as our ineradicable dependence on the folkways of our fathers.

When the adventurer in *Erewhon* was trying to convert his bride to the True Faith, he sought to point out to her the folly of Erewhonian religion. "Can not you understand,'" he said " 'that truth, and goodness, and honesty, and justice are just as real and just as admirable qualities even though we do not imagine them as having a personal existence in the form of rather full-busted ladies somewhere off in space?'" The young woman pondered this for a time. " 'In that case,'" she said, " 'may we not suppose that all the admirable qualities and all the virtues are quite real and permanent without supposing them personified all together in the person of one elderly full-bearded gentleman, off somewhere in space?' At this point," says the narrative, "I realized the impossibility of making her tender young mind comprehend the true inwardness of the Christian faith."

No one seems ever to have inquired whether religion can exist without morality. What would be the use of such a thing?

CHAPTER VIII

PIE IN THE SKY

No ASPECT of our morality is more beautiful than that which pictures happiness here and hereafter as a reward of virtue. Yet, strange as it may seem, this very idea has occasionally been treated with contempt and derision. A slave morality, it has been called, a religion based on bribery or not even honesty bribery—soul-staggering deceit. Of course such vituperation is very silly. It proceeds from a fundamental misunderstanding of morality, civilization and even human nature itself. But it has nevertheless an intelligible basis. Morality is, to say the least, somewhat disingenuous in its theory of rewards.

Our theory of the reward of virtue is that the reward follows righteousness automatically. This idea has never been more happily expressed than in the literature of capitalism. Capitalism is based on capital accumulation; and the theory of capital accumulation is based on the supposition of thrift. Capital is the result of "saving." The thrifty man is he who foregoes present satisfactions for the sake of more remote advantages; and these remote advantages are the direct reward of his thrift and self-control, virtues so highly meritorious that compound interest even unto the third and fourth generations is not too great an honorarium. Since thrift involves stern self-control, it is the very pattern of all the virtues and the very pinnacle of righteousness.

Thus capitalism can be described as a mundane form of the highest piety—and was, about a century ago, by an eminent theologian. This inspired divine took his text from the well-known verses of the Sermon on the Mount, in which righteousness is represented as heavenly capital accumulation. "Lay not up for yourselves treasures upon earth, where moth and rust doth corrupt, and where thieves break through and steal:"—so this sublime appeal to prudence has been worded—"But lay up for yourselves treasures in heaven where neither moth nor dust doth corrupt, and where thieves do not break through nor steal." From this it seems evident that the most thrifty capitalist is he who invests in Heaven, Limited. Like other capitalists, he can do this only by the exercise of some self-control. He may be called upon to forego some of the dissipations of this earth; but his securities are safe and his title guaranteed.

Upon the surface this seems to be an excellent case for circumspection. But it has been subjected to damning criticism, mostly, it must be admitted, by radicals. These radicals hold the capitalist system in contempt for what they insist is deliberate, intentional, Machiavellian inequality. They declare that in its very show of freedom and equality our morality is in the highest degree sinister. As Anatole France remarked, with his wonted irony, "The law, in its majestic equality, forbids the rich as well as the poor to beg in the streets, to sleep under bridges, and to steal bread."

Turning specifically to capital accumulation, they say that it is not the thrift of the poor man but the colossal incomes of the rich which feed the reservoirs of wealth. These reservoirs are accumulated out of the undivided surplus of great corporations and the unspent residues of incomes so swollen that not even the extravagances of a leisure class can sluice them quite away. They even produce many facts and high authority to support this claim, for instance, a secretary of the treasury who reduced the surtaxes on the incomes of the very rich because, as he said, those incomes are the great source of capital accumulation. If the steady thrift of the multitude of poor results in capital formation, say these radicals, it must be somebody else's capital. Each successive census proclaims in figures which even he who runs may read that the poor are becoming perpetually poorer and the rich perpetually richer. Thrift is indeed rewarded—but not the thrifty!

Then, turning to Heaven, Limited, these critics proceed to point out that the rewards of heaven are very hypothetical. They feel bitter about this. All civilizations, they say, have been founded upon some sort of power-system; and in each the power-system has actually rested on the consent of the governed who are always so numerous that they could not possibly be held "in that state of life unto which it has pleased God to call" them without shackles of supernatural toughness. These bonds, in each civilization—so runs the diatribe—are the shackles of superstition, cunningly forged by the enlightened few for the enslavement of the illiterate many. The bribe of heaven, they say, is inequality's most potent weapon. In the darkest hour of deprivation and in the lowest depths of misery it offers the consolation of perpetual hope; and it costs—just nothing. Perhaps the most cynical Hymn of Hate of modern times is the I. W. W. song, written by the great revolutionary poet, Joe Hill, who was convicted of murder and executed during our recent anti-syndicalist picnic under circumstances so dubious that even the president of the United States appealed in his behalf for clemency—in vain: a song with this recurring anthem:

> Work and pray; live on hay.
> You'll get pie in the sky when you die.

But whatever the facts of inequality may be, the case for a conspiracy is very poor indeed. True, our defense of capitalism is not always free from guile. It seems, upon a careful examination of the facts, that our large incomes were not wholly withdrawn from capital accumulation in the days when they were taxed, the difference being chiefly that by virtue of the tax part of the capital accrued to the government whereas after it is removed that portion accrues to the private owners of the incomes. Moreover, in reading the literature of capitalism in which the perfect equity of thrift and compound interest is so glowingly set forth, we are sometimes beset with doubts whether this literature may not be, to some degree, and of course unconsciously, apologetic. It does, sometimes, produce the effect of severe criticism skilfully warded off.

Our enthusiasm for thrift among the poor, too, has inconvenient corollaries. Is it an offset to humanitarianism? We are a squeamish race, and do not enjoy the sight of squalor even on the part of our inferiors. When they were wholly under our charge, in the days of serfdom, we could assist them to provide against total indigence by acting as their trustees and holding back a suitable portion of their earnings for their own good. That is no longer done. Consequently we must teach and encourage them to practise thrift for themselves—or have them on the poor rates; and teaching thrift, upon the whole, is cheaper.

It is true, too, that while no one of us doubts the real value of true religion to all manner of men, we have found the spirit of religion peculiarly serviceable in certain exigencies. Missionaries, we have discovered, perform a very great service not only to religion but to the whole of civilization in their ministrations among the "natives." They teach not only piety and righteousness but obedience and prudence. By bringing the heathen to a proper sense of shame they also create a market for cotton cloth; and since cloth can not be obtained without money, they even dispose the savage to labor on our plantations. So, also, at home we have found that open shops flourish best in the spirit of Christian brotherhood; and that the most effective counter-irritant to labor unrest is a well-conducted revival.

But these are afterthoughts. Was our morality, or our religion, contrived in this fashion? Certainly not. The science of anthropology, obscure as it is about so many things, is clear and positive at least upon this point. Morality and religion grow. Their characteristic forms develop gradually, so gradually as to obscure their dim beginnings and confuse their winding trails almost beyond the power of historians and anthropologists to trace them back. But even if this were not the case, even if we knew nothing of history, and were obliged to rely wholly upon our understanding of the present, we should still know that morality is not a conspiracy against the lower classes. Conspiracy on so large a scale is a psychological impossibility.

In law an axiom prevails to the effect that conspiracy of all crimes is the most difficult to prove. The outcome of various events may suggest collusion. Certain parties may benefit so flagrantly at the expense of others that we can not help suspecting their motives. Real conspiracies are naturally secret; and where they exist the incriminating documents are carefully suppressed. But on how large a scale can evidence be destroyed? How large can a committee be and remain absolutely secure against the indiscretion of its members? Wherever two or three are gathered together, there is dissent in the midst of them. Conspiracies involving more than ten persons are all but impossible. Occasionally, where party lines are sharply drawn and passions are at white heat, wide-spread collusion has been successfully attempted. Half a dozen Indian tribes send around the notched stick and go on the war-path simultaneously. But these are very special cases, where a gulf of language also separates the enemies; and even then the conspiracy can safely be drawn out only for a few days at the most. Given more time, some one is certain to prove too garrulous. Alarmists talk of a continuous, world-wide conspiracy of Jews, or Masons, or Catholics, or Mohammedans, or Bolsheviks. But continuous, world-wide conspiracy is utterly preposterous. The reluctance of our State Department to divulge its secrets to our own Senate, the sworn guardians of the common weal, is sufficient comment upon the theory of conspiracy. No oath has ever been devised strong enough to prevent men from talking too much. When they refer to the supposed "conspiracy of the leisure class," radicals are paying their rulers the highest possible compliment. Any body of men capable of such astuteness and such discretion would deserve to rule!

But even apart from its possibility, a conspiracy to be successful must have the power of achieving its objective. This means that it must aim to produce an effect which would otherwise not come to pass. Is social inequality the result of conspiracy? Then the heavens themselves have conspired against us. On this planet, as it happens, the oceans far outbulk the continents, which is an unfair discrimination against land animals at the very outset. Robert Ingersoll declared that it proves that God preferred fishes to folks. Does it, indeed? According to Malthus' principle of population and Ricardo's "iron law" of wages, the misery of the multitude arises from their uncontrollable spawning. Then the procreative instinct and the whole system of reproduction is in conspiracy against us. We appear to live in a hostile universe!

The fact is that social inequality—whatever may be its causes and its origins—is quite as old and quite as continuous as morality itself. It has not been produced by any moral code, nor is there any likelihood of its disappearing as a result of any merely moral changes. Morality is a human device; whereas inequality and suffering and even ultimate extinction are features of the life of all animal species. Morality is not the cause of inequality; it is the expedient by which we get along with it.

We are a timid and self-conscious species. The difficulties of our life annoy us; and it is to morality and religion that we turn in our hour of need. We are chicken-hearted. If our upper classes had the spunk of bees, they would ruthlessly exterminate us as soon as we had ceased to be of service. If our lower classes were as self-respecting as the wolves, they would put their rulers down and eat them at the first sign of tyranny. But the sad fact is we are not bees or wolves; we are creatures who require solace.

Now the best solace is the unanimous example of our fellows: and that is what morality provides for us. That, indeed, is what morality is: the lock-step of the whole regiment of mankind marching through the valley of the shadow of death with religion playing a cheery tune. This is what Kant meant when he named universality as the necessary form of all morality. It is the essence of the case. A moral act is what all men are doing. When we behave as every one behaves, we are doing our duty. How could a poor man go home without a loaf of bread and face his starving children when he might break a window and steal a loaf and at least make a fight for it, were it not for the "moral support" of the whole regiment of society? Other men are not breaking windows; they are letting their children starve. Society could not exist if misery led men to break windows. It is better, therefore, that a few individuals should perish of hunger than that the lock-step of the race should be broken—with consequences which no one can foresee. So fortified, a man can bear even extreme deprivation with the calm fortitude of righteousness.

These blessings are evenly distributed. The rich are not so callous as we often think. Indeed, are they not by virtue of their culture and their ease the most sensitive of all? It is hard for them to endure the sight of suffering humanity. If they could see and fully comprehend the squalor in which a large part of mankind habitually live, their equanimity would be sadly shaken.

But they too enjoy the consolation of morality and the equity of universal law. They can give ten cents for a loaf of bread and ten thousand dollars for a sable coat with this reassurance, that ten cents is the standard price. They have done what all society must do. Any other practise is inconceivable.

They may regret that some men even lack ten cents. But their regret is quite impersonal. Life is very hard for some people; but it is not our fault. We have kept all the commandments from our youth up; and if every one were to sell all that he has and to give it to the poor—what would become of culture? A hundred million dollars, distributed among a hundred million people, affords just one dollar per head. Our duty is to preserve civilization which, however hard it may bear on certain individuals, is always for the greatest good of the greatest number.

There is vast consolation in the greatest good of the greatest number, precisely the same consolation we derive from mansions in the sky. It is a utopian expression; and all utopias are variants of heaven. The "greatest number" is not a figure that can be calculated. The utilitarians, who brought this expression into use, had a conundrum in which they imagined ninety-nine people in a position to benefit by the extinction of the hundredth man. The question then arose whether the ninety-nine should not push the last, lone, forlorn soul, from whom they all stood to gain so much, quite over the cliff. Would that not be the greater good of the greater number? Such puzzles are absurd. As well suppose that one man stood to gain greatly by the extinction of ninety-nine—a thing much more likely to occur. Would it then be for the greater good of the greater number that all should drift along in moderate discomfort, or that one should attain great bliss at the expense of moderate and tolerable suffering by the ninety-nine? Who can say? We can repeat "the greatest good of the greatest number" over and over until we have hypnotized ourselves without ever extracting the slightest meaning from the phrase. But it is still consoling.

What Beulah Land was to the religion of Jehovah and his Chosen People, the Common Good is to the modern gospel of Humanity (and the Nordic Race). The greatest good of the greatest number is no more a statistical expression than Beulah Land is geographical, and when we apply such tests we only serve to make ourselves ridiculous. Each is a recompense for suffering to those who, in favor of the great reward hereafter, accept the present as it is; a promise of perfection in the sweet by and by. If the promissory note is never paid, what of it? It is not payable in this life, and death cancels all obligations. Civilization does business on credit. With each new generation the credit is renewed. New securities are issued, with new premiums. What does it matter that no cash payment is ever made? The mores are concerned only in financing the present.

CHAPTER IX

MORALS FOLLOW THE FLAG

THE theory that might makes right is always vigorously combatted. The weak deny it because they feel it is no compliment. The strong deny it because it produces an unfortunate impression and is bad for business. Nevertheless, there is something in it.

The counter-argument that, on the contrary, right makes might is obviously true; but it is not really a contradiction and by no means disposes of the fact that might is the dominating element in the situation. Suppose we assume, as the virtuous would have us do, that the mighty are mighty simply and solely because of their superior rectitude. The strong will agree with us, albeit somewhat disingenuously. The lion is mighty in his virtues; but his are not the virtues of the lamb. The point is that after the lamb has been devoured it becomes clear that nature was on the side of the lion. He has behaved as a righteous lion should, maintaining discipline in his domain. We are informed that the well-known couple returned from their ride with the lady inside and the smile on the face of the tiger. The last laugh is to the strong. God is on the side of the heaviest guns. The victorious party has never been known to be in the wrong: it survives to write the history. The first act required of the vanquished is to sign an acknowledgment of war guilt: hence the certainty that right invariably triumphs. Might defines right.

This is the principle which we see at work in the diffusion of cultures from one civilization to another. No people is disposed on first contact to commence copying the moral attitudes of impinging foreigners. The mores are the "fundamental" principles of herd behavior. They are ancient and honorable. People do not abandon what they conceive to be their fundamental traits without a struggle. But without intending any harm to the principles of their forefathers, they see nothing wrong in copying the tools and the weapons of their neighbors—especially if those contrivances seem to be the basis of their neighbors' potency in the arts of peace and war. Techniques are changed first; mores follow surreptitiously.

The "modernization" of Germany was achieved in one generation by administrative fiat of those dominating statesmen who created an empire out of the chaos which Napoleon left in central Europe. Up to a certain point the empire was created by medieval methods. But those methods were effective only within limits. They were effective within the zone of medievalism. Outside that zone stood a great power, overshadowing Europe, whose strength was derived from other sources: from modern industry, in short, the source of British financial dominance and the foundation of British naval superiority. To contest the world with Great Britain it was necessary, for instance, to build a modern navy; to create a modern navy it was necessary to construct an industrial system. This, therefore, the German imperial statesmen proposed to do. Thus the industrialization of Germany transpired not by gradual unconscious seepage but by deliberate borrowing.

But it was no part of the intentions of the

German statesmen to borrow the British social system or the Englishman's ideas of rectitude— known on the Continent as "British hypocrisy." As Veblen has pointed out in his masterly essay *On the Merits of Borrowing,* the Germans improved on British industrialism. By transplanting the machinery they intended not to reconstruct their social system but to add to the orderly and loyal German patriarchy the efficiency of industrialized Great Britain. By this maneuver they hoped to make Germany thrice efficient. A tremendous mechanical equipment, all new from the ground up with no hampering semi-obsolescent units, was to be manned by a labor supply of feudal serfs, trained for centuries to instant, unquestioning obedience with no taint of the radical disaffection which was then already so acute in England. The result was "German efficiency," a temporary phenomenon, no doubt, but important in its time.

The transience of this unearned increment of borrowing is due to the fact that even in Germany like causes produce like effects. To the confusion of the elder statesmen, the German serf did not long remain in his pristine state of unsullied servitude. On the contrary, he became a class-conscious proletarian and developed social democracy with extraordinary speed and vehemence. This is how cultural diffusion works. Like circumstances produce like mores. If social democracy has been the effect of industrialization in one country, the same course is pretty certain to be run in other countries. Germany does not borrow the spirit and morals of industrialized England; it develops in the same direction on its own account. Capitalism mitigated by socialism (or labor troubles) is diffused not by propaganda and direct imitation but by machinery. It is impossible to borrow without borrowing trouble.

Dominance is the slope down which diffusion flows. "Civilize 'em with a krag"—in one form or another this is one of the most fundamental of moral principles. The white man, traveling to the four corners of the globe, everywhere maintains the prestige of his superior clay not by the refulgence of his rectitude but by the caliber of his carbine. Such power is exerted in many ways. A few wanderers may be cast up on a foreign shore possessed of nothing but the training of their hands and such formulas as they are able to remember. But if the formulas work, they will be sufficient. The newcomers somehow dominate and introduce forthwith their new techniques. The Connecticut Yankee, pitched back through the centuries to King Arthur's court, saved himself from immediate extinction by reckoning back to an eclipse which he then proceeded to conjure up. This is precisely the principle which the modern missionary invokes when he precedes his preaching with a little aseptic surgery. Nothing induces men to revise their moral habits so speedily as a first-rate miracle.

Such miracles occur on a national scale. The discovery of America is resulting from just such a miracle. Here again the formula is industrialization; but here it calls for the industrialization of a continent, ultimately, no doubt, of a hemisphere. The ingredients are tremendous resources of coal, iron and the other industrial materials in a land practically unpopulated, and a colonizing man-power drawn almost exclusively from partially industrialized middle classes. India is rich in resources and has been colonized by middle-class English; but the dense population of India has made of those English a feudal caste. Latin America is rich and not densely populated; but it was colonized by feudal conquistadors almost virginally innocent of industry and commerce. Australia, empty and English, is poorer, remoter and later, though in spite of this present handicap it presents the closest parallel to North America and serves to vindicate the formula.

Observing this scene with the omniscience of hindsight, it is not difficult for us to see that the ingredients have been brewing in the pot for a considerable time. The might of America has been apparent to Americans for some years. But only recently has it developed that America is right. The discovery has come as a surprise to Americans as well as to Europeans. We have always been supposed, even by ourselves, to be a sordidly commercial people, bent only upon the dollar, worshiping wealth, indifferent to the finer feelings and the finer arts. This now appears to be a mistake. Having become far and away the richest people in history, we now appear to have been pursuing Destiny. We are so rich that we alone really despise mere wealth. We are so rich that everybody cheats us, so rich that every other people hates us and fawns on us and scrambles to imitate us. In this fashion we have achieved Culture.

One of the chief industries of the world is entertaining traveling Americans. In every capital of the world a show is maintained for our benefit; and where the tastes of Americans are so scrupulously studied it is only natural that American taste should take effect. Thus all the capitals of the world are partially Americanized. Furthermore, American money, American methods and American materials are exported not only in the field of automotive machinery, but in the field of sanitation, where American standards of "decency" lead the world; and not only in plumbing fixtures but in journalism, in phonography, in cinematography, in radiography and so in popular speech and minstrelsy. The shot heard 'round the world to-day is the American wisecrack.

Classicists in all lands and most of all in our own deplore these facts. But classicists are custodians of the past, not of the future. Men of antique culture have their eyes fixed upon the cultural traditions which we once derived from Europe, not on the culture which Europe now derives from us. Is the latter no less genuine? That will remain for the future to determine. For the present it has every mark of authenticity. American popular speech, music and graphic art is, to be sure, vulgar. All folk-lore is vulgar. But like all real folk-lore it is the delight of the sophisticated; and by becoming the delight of the sophisticated—by attaining to literary recognition as "the rich stream of popular imagery"— it has become authentic folk-lore. European musicians are deriving their themes from our phonograph records, as Beethoven and Brahms took theirs from the cheap tunes of vulgar and authentic European peasants.

Moreover, to the confusion of the classicists, we are developing a literary past and a sophisticated present. Not merely are we turning to the arts for "the self-expression befitting a great nation"; we are discovering our own genealogy as the great are wont to do. Each successive issue of our literary periodicals recalls to our attention the neglected greatness of some forgotten writer of our past—neglected not for lack of merit but on account of the deep European shadow in which we were then groping. Light having dawned upon this continent, we perceive our own legendary giants. Series de luxe are being published to document our indigenous traditions. Meanwhile our young people are carrying on the (hitherto unseen) torch, holding aloft the brilliance of America in all the arts. It is no longer *de rigueur* to go abroad to study—anything. New York is the cultural capital of the world. A native American opera has at last fluttered all the critics; and native American song-birds infest that transcendental dove-cot, the Metropolitan Opera House. To be accepted in America is the *grand prix* in all the arts.

Europeans often indignantly protest that whatever our wealth and our resources, our civilization is obviously borrowed. Their indignation is genuine enough: they are being run away with, and they know it and resent it—much as the latter-day Greeks resented the appropriation of their civilization by the Romans. And, as was also the case with the Greeks, the facts are beyond dispute. Rome was a mere colony when Greece was greatest. America has ceased being a colony of Europe only by outstripping Europe. But these facts are irrelevant. All civilizations are borrowed, if we rate them by their sources. With civilization, as with wealth, possession is nine points of the law. The late European civilization is our civilization now.

To the charge that we are debasing and corrupting it we can reply only by shrugging our shoulders. In noting these facts of the diffusion of European culture we have not meant to boast. We need not even suppose that the possession or domination of European civilization is anything to boast about, any more than any like movement at any other time and place. The same charge of debasing and corrupting was made against Rome. But wherein lies the greatness of a civilization? Is it not a matter of prestige? And can it truthfully be said that Rome diminished the prestige of Greek culture, or that America is damaging the prestige of European culture by collecting the glory that was Europe and enshrining it in the grandeur that is America?

The glory that was Greece is a matter of taste

and opinion. Neither Hindus nor Chinese feel that way about it, only ourselves; and we are undeniably influenced by our own (European) past. To a considerable extent it is a Roman past. The language of the Romans was the literary language of all Europe until only the other day; at core the Christian church is still the Roman church. Through the centuries we have inherited the Roman attitude toward Greece. The reputation of Greek culture may be said to rest on that fact alone. When we apostrophize it in superlatives we are only speaking Latin. We say, "Thales was the greatest thinker of all times." We are not familiar with thinkers who wrote in Sanskrit or Chinese; and since none of Thales' works survived, we could scarcely know less than we do of him. But he obtained a certain reputation—among the later Greeks and so among the Romans; and we have never stopped repeating what they said. Is this derogation?

There is no country of the world where Shakespeare, Goethe and Marcel Proust are held in higher honor than in America. Is this derogation? The Greeks believed that by virtue of the middle voice and the second aorist tense, so tragically dispensed with in Latin, Greek was the subtler and more expressive language; and the Romans, being mere centurions and proconsuls, believed this too and have made all subsequent generations of Europeans believe it in the face of perfectly clear philological evidence to the effect that every language is just as subtle as it needs to be. The French believe that by the liberal use of the personal pronoun and *je ne sais quoi* French has become the most finely sensitive of modern languages; and the Americans, mere tinkers and pedlars that they are, believe it too; and the time may come when the dead tradition of the glory that was France will be passed on to the heathen of New China through the adoration of cherishing Americans.

It is said that one can not make a quarrel. Neither can one people make a grand tradition. For that, there must be at least two: one to recite and another to applaud. The grand tradition of Greece was a joint product of Greeks and applauding colonial Romans. The grand tradition of Europe may conceivably go down in history as a joint product of waning Europe and all-conquering, colonial America. Thus like the Romans we shall impose our language, our modern literature, our customs and our morals on the barbarians; but we shall send our sons to Europe to receive the old-world polish. Cycle and epicycle.

But this is only illustration, not prophecy. The actual event may turn out quite otherwise in its details. What is important is the process by which the mores are transmitted and sustained. They are transmitted by the dominant peoples; and they are sustained by the worldly resources which occasion their material success. It may be that in the struggle for existence peoples are sometimes sustained by purely moral qualities. But this is usually ascertained after the event. During the contest all thoughts are centered upon bigger guns.

CHAPTER X

GIDDY EMINENCE

ALL civilizations think well of themselves; that is what it means to be civilized. Righteousness vaunteth itself and is puffed up. A humble and deferential moral code, willing to make allowances for other people's scruples and eager to concede the likelihood of its sometimes being in the wrong, would be a freak of nature, an impossible monstrosity. But no civilization seems ever to have thought quite so well of itself as ours. The reason for this overweening pride is that ours is a self-made civilization. It is pride of origin. Other peoples, with nothing else to distinguish them from their neighbors, have invested all their local pride in the hypothesis of ancestry. They are descended from the gods. Or they are the chosen people. Far back in hoary antiquity Jehovah or one of his equally obliging colleagues has made a covenant with the Common Ancestor—"and thy seed after thee in their generations for an everlasting covenant, to be a God unto thee." But ours is a materialistic age. We have come into prominence after the close of the age of miracles and revelations. Our origins are humble; therein lies our greatness.

Like a self-made man, modern civilization is proud of its humble origins and loves to boast of its reliance on its own expedients. Not that we attribute our superiority to our mere vulgar size and wealth any more than a self-made merchant would do. The self-made man loves to reflect that his affluence and power are the index of his own inward and personal superiority—a spiritual superiority, of course. So it is with a civilization. Our modern culture is of a very imposing magnitude. But its magnitude, so we think, is merely the index of our spiritual superiority.

What are the graces of the soul which lead to empires such as the twentieth century reveals? If we refuse to acknowledge the machine technology, it is clear that we shall have to select for our proud boast its spiritual synonyms. This is very easy. We do not wish to give railways and telegraphs the credit, so we simply say that a superior capacity for government has made possible the linking of a great people. We do not wish to say that modern industry has imposed common interests upon huge populations, so we say that by the perception of their common interests great peoples have become industrialized. Above all things we refuse to think that modern literacy, modern intelligence, "the freeing of the modern mind" are in any way incidental or derivative. So we assert the contrary. We point to the wane of superstition, to the development of "the scientific habit of thought," to the emancipation of the masses from illiteracy. This, we say, is our real achievement. This is the true index of our character. Here are the qualities from which our greatness springs. Behold our genuine moral superiority!

But this connection between our qualities and our achievements is too suspiciously close. Whence came this sudden capacity for letters? Since we are flesh of the flesh and bone of the bone of our totally unlettered forebears, we are obliged to admit that if we are capable of literacy, they were capable of literacy too; if we are literate, whereas they were not, it seems that our circumstances must be different.

Where literacy is concerned our circumstances are different in two reciprocal respects. Machine technology makes the diffusion of letters possible by cheapening the printed page. Nor were our untutored forebears confronted on every hand with the magic symbols by which our destinies are guided: "Subway Entrance, Down Town"; "South Ferry Local"; "No Admittance"; "Employees' Entrance"; "No Smoking"; "Paymaster's Office"; "Line Forms on the Right." Modern technology not only provides the means of literacy, it provides the irresistible incentive. Suppose some stalwart protagonist of the modern soul were to undertake to conduct his business without any reference to any reading matter whatsoever: without signs, without letters, without catalogues and circulars and invoices. He would soon discover why the modern world is literate.

It still remains possible, of course, even though we admit that literacy has been forced upon the world by machine technology, that the human spirit has nevertheless been freed and lifted up by this new, wonderful capacity to read—possible but not certain. We have, to be sure, become literate enough to get about. We have learned the ways of the machine. Our thinking is "scientific" to this extent: we can identify a knock in a misconducting motor. Even the meanest of us can distinguish the angry bark of a burned-out bearing from the inconsequential yapping of a loose wrist-pin. The question is—How does this sort of scholarship affect our other thinking? What good does it do us to be so mechanical, beyond covering the face of the earth with second-hand automobiles? If familiarity with machinery leads people to think for themselves, why is it that people never do think for themselves—especially the people who make the wheels go 'round? Because Marxian socialism reasons in terms of a mechanistic theory of production, Veblen once wrote that it should make a peculiarly poignant appeal to the masses of factory workers whose minds, one might assume, are especially tuned to mechanical ideas. But has it? Compared with the machine technology the spread of socialism has been almost imperceptible; and Russia, where it has taken its strongest hold, is the least industrialized of European countries!

We know that machines do not teach people to think, nor inculcate good taste or good manners or good morals or any of the other paraphernalia of a superior civilization. There are two excellent reasons why they can not and will not. In the first place, there is no peculiar, mechanical manner of thinking—nor even a peculiar scientific manner of thinking. Science is many, not one. Machinery is infinitely varied. A man may know all about one set of machines and be a complete novice where another is concerned, precisely as a miner is just as liable as any one else to get lost in a mine with which he is unfamiliar. Beyond half a dozen platitudes about "facts" and "observation" there is no such thing as a universal scientific method. Each research has its own methods to which the student requires to be trained by intensive specialization. A physicist is as impotent as a theologian in a biological laboratory. If scientific training or mechanical skill were to produce a special type of intellect, there would be as many types of mind as there are types of machinery.

What all the technicians think in common is only the half-dozen platitudes; and they can be mastered in five minutes by a child of ten.

Moreover, even assuming that all machines were built and all researches conducted on the same principles, so that all men trained in science and mechanics enjoyed a common methodology: this would still mean nothing outside the machine-shop, unless we could assume further that the principles could be applied outside the machine-shop. There would have to be a transfer of training from machinery to politics, or from physical experimentation to esthetic taste and moral rectitude. No such transfer is conceivable. A sound mechanical education prepares a man to face mechanical problems; but it prepares him to face nothing else. And the development of a great mechanical technology, with the habituation of great numbers of people to mechanistic processes, prepares a civilization to solve the conundrums of its mechanical technology; but it affords no basis whatever for solving any other problems such as those which confronted civilization before machinery arrived and will remain after machinery has disappeared.

For civilization, as for an individual man, mechanical greatness is mechanical greatness and nothing more. To imagine a civilization superior "spiritually" because of its machines is like expecting Edison to exhibit the qualities of Plato.

There is a notion, in currency among those who should know better as well as among those who could hardly be expected to know better, that science (if not machinery) is elevating civilization by its discovery of "truth." It is nonsense, of course. Science—which is thinking about the mechanical aspect of things—discovers truth about the mechanical aspect of things. It discovers no meaning, cultural or moral. And unfortunately we are greatly interested in meaning. Science tells us how to do things, that is, certain things; but it never tells us what to do, nor takes our cue as to what things to tell us how to do. We have just to accept its offerings of method as they come. Whenever we imagine that it has told us what things mean, we are just mistaken; and great is the folly thereof. The notion that science has "disproved matter" is a conspicuous and recent example of our shame. With the opening up of the interior of the atom, and so on, scientists found that (in certain equations) matter and energy could be substituted for each other. Immediately the cry went up that science had demolished materialism. The world is not matter; it is energy. The soul of God had been laid bare within the orbit of electrons! But then—more recently—it began to develop that in certain other equations matter and energy could not be interchanged so readily. They could not be interchanged at all. So after all, matter exists. What a blow to religion!

What science does not know, the public will not learn. The public does take a great interest in science—at second-hand. The market for the romances, the A. B. C.'s, the Mother Goose Rhymes and the Purple Fairy Books of science is large and constantly increasing. It is becoming the custom to suckle infants upon the wonders of modern astronomy and physics. But this often fails to make scientists of the infants. So far it has even failed to improve the intellectual tone of the public discussion of companionate marriage and the inter-allied debt. To some extent it has given our debate the superficial brilliance of a few long words. The modern vocabulary is scientific. Journalistic hacks writing "true stories" of seduction for the news-stand market employ a physiological terminology that was once limited to doctors. But even this has not made their stories any more edifying than those of Boccaccio and Scheherezade.

One of the great achievements of modern technology is the printing-press (which we imported during the Middle Ages from the heathen Chinese); and it is sometimes argued that through such instrumentalities modern science has transformed our culture. Has it? The prevalence of reading matter is a fact; but what reading matter? After we have made allowance for the relation of literacy to machine technology—after we have admitted that we must read because machinery commands it, and that we can read because machinery permits it—we are still bound to inquire if there is not a margin, over and beyond necessity, over and above the mechanical truths of science, where culture alone holds sway and where the minds of modern men are opened up and "freed." There is such a margin, surely. It is the zone of the tabloid press, the papers printed "for people who think."

Admirers of Lord Northcliffe credit him with the discovery and first exploitation of this area. According to legend, while he was still a young man he came suddenly to the realization that vast hordes of people were being taught to read by the modern public schools who, after they had learned, found nothing to read suitable to their station in life. He set about to satisfy their want; the modern gutter press is the result. This story may give too great honor to the master journalist. Some say he learned of the literary proletariat from his first employer; and others point out independent trends in popular journalism, and the extreme facility with which his ideas were adapted from England to America. In any case, one thing seems certain: the inevitable food of the literary proletariat is literary garbage. Teaching people to read only prepares them to read what there is to read—what is provided within their means.

As to the "superior capacity for government" which has "made possible" the development of machine technology: it is democracy in name and plutocracy in fact. "Self-government" is invisible government—nor is it so very invisible. Now and then—as, for instance, at the present time—the veil is lifted slightly, and we see a group of speculators at a national convention stipulating a presidential candidate who will make cabinet appointments agreeable to them and then, their conditions having been met, financing a national party, paying for a presidential campaign, buying their cabinet officers outright and putting them in their pockets—along with sundry government possessions.

Probably we should not bother to call our government democracy except out of deference to history, just as other countries, equally plutocratic, call theirs monarchy out of deference to a historic past when their kings were something more than fashion mannikins and jockeys. At one time democracy meant government by election and election meant the personal choice—no doubt at the instigation of men of substance; but still actual choice—by a majority of men, all of whom knew each other, in a small group like a New England town meeting. The idea behind the assignment of representatives to small districts—so much condemned in recent years by advanced liberals and other idealists—was that the men of a local district knew each other and could in fact elect. The same idea underlay the original, constitutional method for the selection of senators and presidents. The senators were to be chosen by state legislatures because the state assemblymen might be presumed to know and have a personal opinion about every man in the state of enough political prominence to rank as a possible senator; while the selection of a president was delegated to a college of electors who might similarly be presumed to be able to canvass the field of available candidates.

All this is changed to-day. The local district of a representative has to-day become the basis for the distribution of patronage and the organization of party machines. The election of senators by state legislatures had to be abolished because state legislatures were too patently venal. The constitutional technique for the election of presidents has become a mere rite. Ballots are really cast for the candidates themselves; and the size and variety of the electorate puts the highest possible premium on colorless, compliant men. A striking and powerful personality can become president only by some freak of chance. Transparencies are the rule. "Personalities" manufactured by the press-agents are easiest to manage.

Nevertheless the system by which the country is (ostensibly) governed is still called election. The selection of officials by this roundabout technique of hoodwinking electorates is still called democracy. Therefore it is possible to say that we now conduct upon a continental scale such government as our forefathers projected upon the scale of the New England town meeting, and to make the merely verbal fact the basis of our claim to spiritual growth.

Whatever conceits we make the basis of our boast, our logic remains the same. Take the famous battle-cry of freedom: liberty, equality, fraternity. We can dismiss fraternity at once as sentimental and quixotic. This is the day not of fraternity but of the Nordic race. "Americanism" now means not the preservation of the principles of Thomas Jefferson but the preservation of the vested interests of the descendants of Thomas Jefferson's contemporaries.

We still hear much about liberty and equality. It is a matter of pride with us that ancient slavery and medieval serfdom have disappeared from our midst. They have been replaced by the wage-system of industrial employment and the renter-system of agriculture. Here is a case where the word has been altered while the reality has remained. The small shopkeeper too, who was once the sign and symbol of individual freedom, is now a mere creature of the agencies which control his credit and his goods. Liberty is chiefly remembered in our courts when it becomes necessary to preserve the inalienable right of the laborer to work at night, of the woman worker to work whatever hours please her, and of the child to enter the factory when the whim has seized him (or his parents). Equality is remembered whenever it becomes necessary to reassert the equality before the law of great corporations and individual employees.

The purport of all this is not pessimism. The lower classes are perhaps no worse off in modern industrial civilization than they have been in other systems. They may even be better off. But it is not necessarily self-evident that they are better off. We can not infer it directly. What is directly evident is change. Literacy has spread; but the meaning of literacy has changed. Democracy has spread and Americanism has greatly enlarged; but both have had their faces lifted. Slavery and serfdom have entirely disappeared; but the situation which results is not quite as though all former serfs had been elevated to nobility. In each case industrialism has altered the appearance of things. In each case what has happened has been simply industrialization.

The industrial scene is boringly evident to-day. But it was not always so much in evidence. Considerably less than a hundred years ago it was possible for men to perceive with interest and trepidation the changes industrialism was bringing without being clearly aware of the machine technology behind those changes. They saw the world growing more literate, and they rejoiced. They saw great populations springing up and great empires in the building; and they supposed, not unnaturally, that a new millennium was dawning. Man was becoming a political animal, even a moral animal.

This obvious mistake comes from our incapacity to visualize great change. We see one change, and we immediately proceed to cast up all accounts, forgetting that other changes are in train. People learn to read. Overlooking all the causes of this alteration and all the prospects for more cataclysmic breaks, we at once assume that they are going to be what reading men have always been in the past, namely men of culture and refinement. People learn to vote on a continental scale. Overlooking all the susceptibilities of continental electorates to manipulation, we assume at once that they are going to vote as the House of Commons, or the town of Boston, voted in 1642. Even to-day, when the contrary is as conspicuous as the sun, we still go on attributing to modern civilization wholly inferential virtues: the virtues of reading men and voting men of the good old days of yore—before any but the best could read or vote.

Our achievement is one of mass. Looking back upon our small beginnings brings us a tremendous thrill—like the thrill of achievement which emanates from Benjamin Franklin's biscuit and John D. Rockefeller's dime. But alas, all that we hold precious in our modern greatness is illusory, like the magic of biscuit and dime. It was conceived after the event, like the biscuit and the dime. First one becomes rich and then one reconstructs the moral stamina with which, of course, the riches have been gained. First we perceive the effects of industrialism—large populations, machine methods, universal elementary education; and then we conceive that those effects of industrial development signify real spiritual growth. In our imaginations, flushed with the heady liquor of industrial success, we see the printing industry as culture and the railway and telegraph as democracy.

CHAPTER XI

ONWARD AND UPWARD

WHATEVER the purposes of righteousness may be, they are always attributed to the world at large. The main purpose is the maintenance of a social status quo: to keep those in power who are established in the seats of authority; to repel all possible foreign invaders; and to convert the heathen to the true faith, chastity and the use of bathtubs. For some reason or other these purposes are perpetually in doubt. They seem to be unable to stand on their own merits as civic improvements. Or perhaps it is necessary for man always to imagine the cosmos associated with him in every enterprise. At all events, morality is always accompanied by a moral theory of creation, a righteous geography of the heavens and the earth and the regions under the earth, and an edifying history of the higher anthropoids; in all of which it appears that the powers of light have been engaged in a gigantic conspiracy to produce this very result, to select this very people for ultimate triumph over all the neighbors, and to bring these very rulers to their consummate reward.

In earlier and simpler times these grandiose gestures were accomplished with the pageantry of superstition and in the language of theology. But superstition and theology having passed away, the only materials remaining out of which a cosmic geography and history can be fabricated are the materials of science. Consequently we have been obliged in recent years to couch our theory of cosmic purpose in a scientific language. Fortunately this has been very easy. So edifying indeed is the language of the evolutionists that their vocabulary has quite taken the place of the will of God among modern apologists for righteousness. According to this language the world is not merely undergoing a process of universal change such as every one can see: the process is one in which virtue is rewarded; a process in which excellence invariably triumphs; a process in which the victories of the better over the worse are cumulative: throughout the length and breadth of the cosmos the good gets continually better while the worse disappears into eternal oblivion. Furthermore and most especially, it is a process which culminates in us. Chaos becomes cosmos; the inorganic becomes organic; matter becomes mind; mind becomes soul.

So stated—and it is almost universally so stated at the present time—the whole evolutionary hypothesis is evidently moral, not to say religious. It is the doctrine of universal moral progress. But of course the doctrine is most explicitly moral where it touches human society most explicitly. There it is not only a theory of betterment: it is a theory of the betterment of morals. Not only is this planet a better thing than the primordial star-dust, and man a higher species than the animals: modern civilization is higher and better than earlier cultures and the righteousness of the modern man is a more righteous righteousness than righteousness has ever been before. What are the traits upon which we chiefly pride ourselves? Our reason?—or our social graces? Evolution shows us to be more reasonable and more highly socialized than man has ever been before. All human history is a "rationalizing and socializing" process; and we are the ones who top it off. How can we know that our moral principles are true, sound and righteous altogether? Evolution proves it. Does not the sounder, the fitter, invariably survive, selected by the evolutionary process, the more successful because ineffably adapted to guiding man to his manifest destiny? It does—or so we say.

We say so, alas, not because the data of science prove any such thing. Strange to say, the facts of astronomy, geology, anthropology and all the rest are quite indifferent to the purpose of anybody's mores. But that is what we have always wanted to think; we say it because we want to think it still, and because we can if we wish merely by interpreting the thought into the scientific data. In general there are two forms of interpretation by which we achieve this feat of edification. One serves to show that history is a "rationalizing and socializing" process; the other exhibits modern righteousness as sounder, fitter, more successful because it is later and has survived. One is interpretation by selection and arrangement: we select whatever it is we want to celebrate; of course there are many other things beneath the heavens, but we ignore them and select this; and then we show what went before and what went before that and what went before that, until we have gone back far enough to make it seem that the cosmos has been in labor to produce just this. That is how we prove man the "highest" species, and modern civilization the "highest" culture. The other interpretation is functional and descriptive. We take the process of perpetual change, which is a function of all things whatsoever, and we describe it in such a way that it will seem when we have finished to be not merely change but limitless, progressive betterment. To be sure, the data exhibit only change. But betterment is change; and so if we call the process betterment we shall not contradict a single scientific fact: we shall only add a little something to the facts—a little edifying touch.

In order to prove that modern civilization is the "highest," we must emphasize the fact that it is the latest. In order to prove it the most completely "socialized and rationalized," we must assume that the test of these qualities is size.

That is what they mean: size, large-scale mechanization, literacy and so on. If our culture were the fruit of tiny city-states, we would select other words in order to idealize it. We would say that it is refined, free, sensitive, or something of the kind. But it is big and muddled. Therefore we translate "big" and "muddled" into the language of approval and say "socialized" and "rationalized." Then we show that other cultures have been smaller. This exhibits their deficiency in the quality we adore to possess.

This manner of interpretation was inherited by the sociologists who invented the phrase "socialized and rationalized" from the biologists, who had long since been busy with the demonstration that man is the highest species. They worked with anatomical qualities; but their technique of selection was just the same. Man has a large brain and central nervous system; therefore the biologists selected the central nervous system as their basis of comparison. But central nervous systems are a rare possession among the animals, being limited to a few of the larger species. To insure success the biologists selected size, also, as the basis of comparison. On this basis a continuous line could be run through all creation. Dignify size by calling it "complexity." This seems reasonable. Surely larger things are more complex. Then by a slight stretch of the imagination we can picture the massing of stardust into a planet as an increase of complexity (or at least of density, though some shrinkage occurs) so that the development will appear to be continuous from the simple to the more complex wherever we look in nature. This even provides the sociologists with their precious cue; for since our civilization is the biggest, what easier than to trace the ascending curve of growing complexity straight up to the edifying present?

Unfortunately there are a few difficulties with this interpretation. They are subtle, perhaps, but in the end they appear to be fatal. What is the standard of complexity in science? Suppose we regard the matter in the light of mathematics, as scientists so love to do. Organisms are composed of cells; cells are composed of molecules; molecules are composed of atoms. Once upon a time we imagined that the regress ceased with atoms, which were final and "simple." That simple notion has long since gone by the board. Atoms are now conceived to be indefinitely complex—how complex we have no idea. The hypothesis has been advanced that the atom is a planetary system of electrons and protons, in which case we have no guarantee that there is any finite point—like the old-fashioned atom—at which things stop dividing and are simple and undifferentiated. Let us suppose, then, that something of the kind is a "fact": then the progressus from complexity to simplicity is a regress along a continuum which has no end. We are face to face with infinity in this matter. Now it is one of the leading characteristics of infinity that infinity decreased by any finite sum remains still and always infinity. If the series of degrees of complexity in material arrangements is infinite, then any given material object is infinitely removed from "absolute simplicity," just as each of the fractions ⅛, ⅙, ¼ is infinitely removed from infinitesimal. Precisely so the algæ are infinitely complex, no less so than *Homo sapiens*. In short, we have no basic simple that is a measure of all complexity, as an inch is a measure of all finite distances. If we are to sustain the evolutionary illusion of progressive complexity, we must assume by arbitrary fiat that any two-celled micro-organism is "more infinitely complex" than any two one-celled micro-organisms. Then it will follow that a billion-billion-celled organism is still more infinitely complicated. Even so we should have difficulty in explaining why man is more complicated than the whale; but let that pass for a moment.

We have now reached the point at which we can, presumably, range the creatures of the earth in evolutionary order declaring them to be in order of increasing complexity as they are (in certain instances) in order of increasing size. But here we are faced with the difficulty of the whale. Unfortunately for our picture man is by no means the largest of God's creatures. Consequently, if we are to sustain our thesis that evolution (which reaches its pinnacle in man) is a movement from simple to complex, we must find some other basis of complexity than size. We establish the fact of increasing complexity by an appeal to size. That is how we got the concept of complexity under way in the first place. But somewhere between micro-organisms and man we have got to alter its whole character.

Now it happens that man has certain physical characteristics that are more or less unique. Most species have. In man's case we may mention, for instance, the excess of cerebrum. This is our *deus ex machinâ*. What we mean by complexity is now to be development of cerebrum. (This standard could hardly apply below the vertebrates; but let that pass.) Thus the whale and the elephant and the poor, defunct dinosaurs are big enough in other organs; but by comparison they make a poor display of brain, the dinosaur having more gray matter in his spinal ganglia than in his head. But who selected brain as the standard of comparison: nature, the dinosaurs, or ourselves? Obviously, we. The dinosaur is extinct; but when he flourished we were not yet thought of. (Cro-Magnon man had a larger skull capacity than we; and he is extinct too.) No, if by complexity we mean simply our kind of complexity, we may as well avoid confusion and say outright, dogmatically, that the trend of evolution is toward man, and no nonsense about infinity and dinosaurs!

From this dilemma there is no escape except through sighs and tears. We can arrange the evolutionary series in order of bigness: that is arbitrary and even so leads upward but not to man. We can arrange the series in order of cerebral development: that is equally arbitrary and leads to man but not from micro-organisms. Between the two arbitrary assumptions there is a complete hiatus. As an interpretation of the scientific data the theory is not quite adequate. Such interpretations are derived not from the data of development but from the state of mind of the present generation.

It is no mere accident or coincidence that evolution leaped from obscurity to celebrity in the latter nineteenth century. As an expression of the state of mind of modern man the theory has every perfection of literary form. Man is the greatest of the animals (that is, the most complex). Modern civilization is the greatest of all civilizations. That is, it is the biggest, or at least (for we must not forget India and China—the whales and dinosaurs of history) the most highly concentrated and sharply focused; anyhow the most highly socialized and completely rationalized. That is, the best. Another people would have deified the common ancestor and traced all achievements to his god-injected blood. They would have perverted their history at its source. We have outgrown miraculous sources. Nothing will serve our purpose but the perversion of all history, from beginning to end, ours and everybody else's, in one grand cosmic miracle—the evolution of ourselves.

Not content with our own special theory of the direction of evolutionary history, we must also impose our interpretation upon the inner process. Not only is change a function of all things, living and dead: it proceeds by the survival of the fittest. This phrase, invented by Darwin and repeated by his successors in the faith, has had a most exciting effect first upon sociologists and moralists and later even upon the general public. It has become the gospel of progress. Its effect is typified in the response of Sumner, the great proponent of the mores. It so befuddled him that he proceeded to write on alternating pages of his book that only mores make things right and that mores which have survived in the evolutionary struggle are "effective" mores, "fit" mores, and therefore good mores. Since all have survived, obviously all are good; and since ours have survived latest, obviously ours are best. Sumner is not quite so explicit; but the implication is inescapable.

The superficial contradiction is obvious. On one hand, no standard of value is said to exist save that of custom; on the other, a standard is announced, independent of custom, whereby present customs are held to be better than past customs. The standard must be independent of the mores, because all mores proclaim their own righteousness. They are "by construction," as the geometers say, self-justifying. The existing scheme of mores does of course proclaim its own righteousness and the barbarity of former customs. But former customs did the same, even condemning present customs in advance precisely as the moral standards of the present frown upon all symptoms of unrest and change. The "judgment of history" is quite irrelevant to the claim of any given morality. The two are, in fact, directly opposed. Barring evolution, history passes no judgment on the mores save the inevitable one of decay and dissolution.

Evolution seems to reverse this judgment; but what it actually does is only to raise the limit of the game. Whereas before evolution we thought in terms of individuals, we now think in terms of species. Evolution is no guarantee against extinction even for species. At the very moment when we are vaunting our triumph over nature we may be heading straight for the oblivion that overtook the trilobites and dinosaurs. There is nothing in evolution which in any way guarantees not merely the continuance of man but even the continuation of the "upward" process toward "more complicated" forms. A slight shift in the weather and we may find the planet once more inhabited solely by one-celled plants. We sometimes overlook the fact that in the onward march of evolution one-celled plants have not been left behind, nor neolithic culture. The world is still chiefly inhabited by micro-organisms which—were we to disappear—might be as sublimely unaware of our brief reign as would be Australian black men of Europe's momentary glory. Evolution proclaims the survival of the fittest. But which are the fittest: the one-celled plants, primordial and ever-present; or that hasty afterthought of nature, *Homo sapiens?* Which is the more durable civilization: the one that has been for fifty thousand years the same, or the one which we invented day before yesterday?

The principle of change provides no answer to such questions. Even if we go no further in our analysis of its significance, the fact is already incontrovertible that evolution does not reverse the judgment of history, and that it is no basis whatever for discrimination between supposedly "effective" mores and those which have already reached their limbo.

But it is nevertheless necessary to go further. The whole concept of fitness, with which evolution works, is as misleading and vicious as any figment of the most naive primitive mythology. It is an invocation pure and simple, a verbal spell under which the intellect is numbed and emotion stimulated to run riot. Many contemporary scientists have become aware of this and are cutting down their dosage of this volatile ingredient. Evolution, they say, is not the "survival of the fittest" but the "survival of the fit." Why the change? That calm, monumental, unimpassioned intellect, Charles Darwin, used the superlative degree.

What does this deflation signify? Primarily it

signifies the noxiousness of the evolutionary drug.

If we regard the survival of the fit (or fittest) not as a utopian doxology but as a biological hypothesis, its meaning is very simple and quite clear. It means, as history shows, that in any given period, under any given set of conditions, certain species are becoming extinct and others are flourishing. The biologists have arbitrarily assigned the word "fit" to this situation, describing the lucky as fit and the unlucky as unfit. In this they have followed the lead of Darwin who—contrary to the popular impression—was a man of vivid and dramatic imagination. (Consider the state of mind of a scientist who makes the wooer's choice one of the leading principles of his scientific exposition!) Of course the fit and the unfit may be in reciprocal relation. The last two centuries have been disastrous ones for whales: we have exterminated them. Alas, the whales proved unfit! That is to say, we approved of them so heartily that we overdid our approval. Like the goose that laid the golden egg they were deficient only in not being indestructible. Now that they have disappeared (almost), we commiserate with them in such language as a highwayman might use to the victims of his trade, "Better luck next time, old toff."

In this particular case an added note of irony is supplied by the historical setting of the "struggle for existence." Whales and men lived together on this planet amicably enough for let us say one hundred thousand years. Then a moment occurred when man, having developed an incipient industrial system, town life, the domestic lamp and the necessity for artificial light, still lacked a common lighting fuel. So the whale was attacked for his oil. The attack became absolutely disastrous for the whale when the advance of the industrial arts brought the whaler's tools to their present pitch of perfection. But the same advance made the whale quite unnecessary to our industrial economy. Whale products still have their uses and command their price, of course. But it is impossible to believe that in the present day of the ascendency of oil the fiasco of the whale could be repeated. By the accident of fate, the whale was exterminated by the same expedients which immediately after made the extermination of the whale an unnecessary butchery. But the logic of evolution is impervious to irony. In a succession of glacial and interglacial periods, creatures are killed off in one, three and five who could have done quite nicely in two, four and six. But this, too, makes no difference.

In short, the expression "fit" is meaningless. The survival of the fit is the survival of those which survive, a pure tautology, a mathematical substitution: the term f for the term s. If we inquire why, then, the substitution has been made, the answer is that intellectually the term f means nothing; but emotionally the word fit is no end potent. It stirs us. This is what the biologists have recently perceived, and that is why they have toned down Darwin's "fittest" to the less disturbing fit. They recognize that fit means nothing which they can defend in scientific language; and they say so in all their more erudite publications. But even fit still means too much. Darwin's purple patch has paled a little; but it is still purple.

If any one has any doubts about this let him make another mathematical substitution. Instead of saying that in the struggle for existence the fit survive, let us say that in the struggle for existence the unfit disappear. Let us go the whole hog and say that the least fit disappear first. Strangely enough this form of expression is much closer to the realities of the struggle. When we say, positively, that evolution is the survival of the fittest we seem to project upon the screen of imagination a sort of caucus in which through a series of eliminating ballots, the favorite son is coming to the front. But nature knows nothing of favorite sons. In nature it is the eliminated who are singled out. The balloting is never over. It is a continuous process of weeding.

This is the form in which the theory appears in the work from which both Darwin and Alfred Russell Wallace received the idea—Malthus' *Essay on the Principle of Population*. Malthus was concerned with the process of diminution of populations by war, famine, pestilence and (as an after-thought) moral restraint. It was with decreasement that he was concerned; and it was the subtrahend, not the remainder, which first challenged the imaginations of Wallace and Darwin—the unfit who disappeared.

In the working of this process, so it seemed to them, certain orderly principles of selection might be at work. A cold period, and all who are too thin-skinned must go; a hot spell, and the wooliest ones are straightway parboiled: the disappearance of the unfit.

However far we carry this formula it still holds true. We have described the evolution of mores—the progress of civilization—as a process of natural selection in which the fit, the "effective" culture has survived. Let us say instead that unfit civilizations have disappeared. Is this untrue? Not in the least. So far as the scientific principle extends, it means exactly the same thing: a struggle, a slaughter, the last roll-call.

Emotionally, the effect is of course far different. When we think of the survival of the fittest, we incline unconsciously to the presumption that the fittest are after all pretty good; rather better than mediocre, anyhow; the best of a mighty varied lot. On the other hand, the disappearance of the unfit has a meager sound. It gives the impression of a cliff uncomfortably near over which some of us have already tumbled; a fate that may await us next; the great unknown; life's inevitable end; the undiscovered country from out whose bourn no genera return!

For aught we know, human civilization may be a dubious experiment. Let us not be misled by its apparent age: that feature is only apparent. In the history of geological conditions, in the history of the development of living things, our apparent antiquity is but a moment, a mere intake of nature's breath. In the next moment we may be spewed out—for we are not the sole survivors of the selective process. In the great throng of creatures who, like ourselves, have not yet fallen over the cliff and disappeared, there are many who have weathered heat and cold, humidity and dryness through millions of years. They have seen many of this fancy, experimental species disappear. What reason have we for supposing that they are near the cliff while we are on the far and safer side? Only our righteous self-assurance.

Civilizations, too, have come and gone. History is full of them. We seem now to be at the height of our rather exuberant powers. Is this evidence of longevity? Is it not true, rather, that all civilizations have at some time or other been at the height of their powers, and that their powers were even then proving to be decidedly precarious? Perhaps not fitness but extreme precariousness is the constant quality of all human arrangements. This is not very reassuring. Our arrangements may be slightly less imbecile than theirs—or they may not. We shall discover in the course of time from evolution, perhaps by waking up and finding ourselves extinct.

One form of imbecility in which most of the defunct cultures indulged themselves to deplorable excess is tyranny of the strong over the weak; and one of the favorite techniques of extinction has been the revolt of the weak against the strong. Whenever this occurs and the strong are unceremoniously bundled, bag and cultivated baggage, into oblivion, we say that a civilization has disappeared. What has disappeared is not a people but their rulers and their rulers' mores together with all their various styles in self-expression. With climactic effrontery we deny to such changes the grace of evolution. This, we say, is revolution, not evolution. If anything is still wanting to exhibit what we mean by evolution, this charming irony should do the trick. The extermination of the whale is evolution. The extermination of Indians is evolution. The extermination of Huns is evolution. The extermination of the poor is evolution. But the extermination of the rich, the extermination of the idle, the extermination of the arrogant and cruel—that is not evolution; that is not what nature has intended; that is abominable and revolutionary.

Such is the edifying evolutionary gospel, self-satisfied and righteous. Over against it stand the facts of nature: nature, the arch-revolutionary; nature, red in tooth and claw.

CHAPTER XII

CONTROL BY EMULATION

IN THE main, the control of morality is secure. Social emulation is stronger than the division of the classes. It arises from the very character of human herd behavior. In our domesticated life we proceed by habit and tradition, that is, by doing as others do; and in spite of slight risks here and there, we are a civilization by virtue of that inner necessity which impels us all to look in the same direction for our standards. Class emulation is a condition—the successful condition—of all righteousness.

No doubt it is a very good thing that this is so: a good thing, that is, for civilization and morality. Social emulation is the force which holds every man in his place: the end justifies the means. In *Major Barbara* Shaw has drawn a striking picture of the moral system with each class dovetailed in its place, each with its foot placed firmly on the one below, each with its eyes fixed vainly on the one above, each held by all the others precisely where it is. The wife of the chairman of the Board snubs the wife of the rising young treasurer; and the latter retaliates by employing at a larger fee the same interior decorator who has set the standard of taste for the chairman's family. The wife of the conductor snubs the wife of the brakeman; and the latter retaliates by teaching her son to play the violin.

Viewing this simple and obvious system, philosophers have been moved to declare that it only goes to show how all men "really" hold the same ideals in reverence. But in their usual preoccupation with the abstract the philosophers have lost sight of the concrete. They have grasped the nature of the universal process; but they have quite mistaken the concrete elements which make it go. Their way of visualizing social emulation is to compare a pig to Socrates, or Alcibiades attending a classic tragedy with his lackey sprawling dead drunk outside the theater. The difficulty with this form of statement is simply that Socrates and the pig are not on speaking terms. A miserable man may be better than a contented cow; but cows are contentedly unaware of it.

On the animal level (where men and cows consort as equals) the pleasures and pains of existence are organic states. Each is probably a general organic condition, produced by the suffusion through the body of glandular hormones and so on. Each can be produced by concrete agencies, such as a bale of hay or a bottle of wine. It is possible to argue that these conditions are selective ones organically. By and large, blows are deleterious whereas bales of hay are to the good. Even animals seem to know this. At all events they avoid one and seek the other, a mechanism of pleasure and pain, avoidance and pursuit, that seems to be fundamental for all living things, and can actually be observed in one-celled micro-organisms no less than in Socrates himself. Even plants depend upon it, bending toward the sun and probing their way around the boulders; and if we raise the query whether they enjoy the sunshine and detest the rocks, we shall find ourselves not better able to assert that the cow enjoys her corn, and may even come up with a start against the realization that we have no trustworthy evidence that Alcibiades enjoys his play except that he frequents the theater. Consequently it seems reasonable to infer at least that the same principles are guiding all living things to what is best for them.

But the disconcerting fact is that the cow is still guided to corn in spite of Socrates' refined preference for philosophical discourse, with never an envious glance to prove the superiority of reason over brutishness. We need not allow ourselves to be detained by the question whether the animals are complete materialists. Perhaps they are not. But they appear to be; while not even the most inebriated man subsists by wine alone. He never ceases to be an animal, to be sure. In his most esoteric moments, when he is dealing with the most recondite interests of which he may be capable, he is still avoiding and pursuing, trembling with anguish or exaltation. But brute nature has loosed its grip on his behavior none the less. His organic conditionings are still primitive; but he is no longer seeking mere food and sunshine; he is in pursuit of goods of his own invention. The enjoyment of sunshine and corn is a natural enjoyment for both men and cows because sunshine and corn produce the pleasure-pursuit responses willy-nilly. But is the effect of Æschylus or Demosthenes on their hearers similarly automatic? No, on the contrary it is wholly secondary, wholly mediate and derivative.

This was the point at which the great, if humorless, philosopher, John Stuart Mill, ranged the pig and Socrates alongside the same trough. The principle which this strange maneuver was intended to establish was first that there are higher and lower pleasures (as every one admits) and, second, that every one who has had experience of both prefers the higher. It may seem strange, perhaps, that he should have selected such a couple. Whether the pig was to try philosophy or Socrates the garbage he quite failed to specify. The reason for this highly strategic omission is that he was bound to escape the implications of his own principles; for without doubt the pleasures he had in mind as best were not the superiority of champagne over beer, but the alleged superiority of all delights of intellect over any conceivable indulgence of the senses. This is a plain case of compensation mechanism. The thinker, excluded for ever from the simple enjoyments of the multitude of morons, assures himself that he would not be a publican or sinner if he could, and imagines that all the publicans and sinners would straightway abandon their infantile amusements if they could but catch a whiff of the nectar that he quaffs. "Wisdom is better than strength," saith the Preacher. But was he a robust fellow? Mr. Bernard Shaw announces (some years after writing *Major Barbara*) that the intellectual ecstasy of the octogenarian is more vivid than the sensual delights of youth; but Mr. Shaw will never be young again, if he ever was especially gifted as a connoisseur of the pleasures of the flesh, which is doubtful. Mill's case is not even doubtful. A more bloodless man probably never existed. Let us not be deceived by his casuistries of compensation; intellectual ecstasy is not what the public wants.

Not brutish debasement but envy is the key to the more elevated pleasures. Mill was right about the principle and wrong only in his concrete interpretation of what the principle must mean. Pigs do not sample philosophy. Drunken lackeys do not admire Æschylus. But there are pleasures which even the lowest covet, namely, more and better beer, actual adventures instead of moving pictures, one's own automobile instead of a taxicab; a Rolls instead of a Flivver. Here is the automatic principle, running through all the pleasures and pains of life. If it leads in a different direction from the abstract visions of philosophers, that is the lookout of the philosophers. When we are going after concrete facts we have to take what offers; and the concrete fact is that all people do prefer the higher pleasures as they conceive them, and they conceive them after the pattern of class-stratified society. These are not animal fulfillments. They are the complicated satisfactions of domesticated men. They are learned satisfactions; and they are learned in the school of civilized society. Hence they are really moral, as no mere animal taste can ever be.

But what we learn from civilized society is that some people are better off than others, and that it is infinitely good to be better off. We also learn who the fortunate ones are. They are not philosophers. For the general run of people, great and small, well-to-do and not so well, not only are philosophers, thinkers, intellectuals quite beneath contempt, they are completely out of sight. As the automatic principle of social emulation works, it is the great ones of the world who are held up to universal envy and universal emulation—the rich, the strong and powerful: the rulers, owners and exploiters of "the greatest number." These are the people whose taste dominates all taste, whose pleasures are desired and imitated, or at least witnessed with eager appreciation from afar by the miserable multitude. We have no common measure between a discontented saint and a contended cow; but we have a common measure of all the strata of society: they are all looking toward the top.

This is the principle which Veblen called "pecuniary emulation." But the emulation is not pecuniary; it is moral. In a pecuniary society it takes a pecuniary form. Power and leisure and an amplitude of worldly goods are all signified by a million dollars. Therefore wealth is the convenient shorthand of our civilization. It is easier to say merely that we want a million dollars—that is, more and better of everything—than to specify in detail.

But if we were to specify, we should find ourselves specifying the emoluments of wealth and power. I may say that I would like to have the leisure for extended study; I would like to study without regard to the professional uses to which my studies might be put. The philosopher's ideal? Perhaps. But literally translated it means that I would like to be rich. Suppose an alternative were offered: on one side the opportunity to study for ever, in any direction and without the slightest pressure except that I must study incessantly and without surcease; on the other, freedom to study or not, as I please. Which is the more attractive? In one case, by a very slight turn of the wheel of fate,—say, affecting health,—I find myself a slave; in the other I am assured of freedom.

But—lest we imagine that freedom is the ultimate of happiness—the freedom which I crave is the freedom which the wealthy seem to have secured; and in so far as I am envied by those who suppose themselves inferior, it is for the freedom (and other perquisites) which I seem to have secured. These appearances may be illusory. That makes no difference. We trade on illusions. All enjoyments above the level of the cow may be illusions. The cow would think so. But they are the enticements that rule us none the less. Every now and then some reformer rises in dismay to cry out against the perversity of the poor who prefer the illusion of decency (as they conceive it) to the solid (but animal) comfort of good bread and butter. What does this mean but that even the poor are men? Even the poor conceive themselves (however falsely) to be better than contented cows; and their divine discontent takes this form: that at the cost of an actual pinch in the pit of the stomach they will put up a bold front in public, flaunting gay silks over tattered cotton underwear. It is impossible to detach the poor from envy of the rich because it is impossible to detach the rich from self-satisfaction.

By preferring higher things, moreover, each class completely realizes its place. The major business of the mores is to regulate the behavior appropriate to caste. We are speaking of the mores which express a whole civilization by knitting all classes together as a whole. Inside such a system there are always subdivisions, semi-detached systems of righteousness countenanced only within the ranks of a given caste. Such, for example, are the morals of business. Many people have wondered in recent years why the community of substantial, upright business men have been so strangely silent with regard to the behavior of certain of their fellows accidentally revealed by prying politicians. A certain millionaire while engaged in the conduct of his business, which is oil, purchases a national administration in the open market, subsidizing an entire national party for insurance; and when he is caught employs a private detective bureau to corrupt a jury. But the reason for the general silence is that such practises are general. They are part of the special mores of business men. In commerce it is customary to employ spies, especially when labor troubles threaten; and luring valuable employees, even executives, from their loyalty to their former employers is *de rigueur*. Understand only that a business man regards government as another business (just as he regards business as a branch of government) and the case becomes quite clear. In a similar fashion sabotage and riot are the common and inevitable mores of working men, always understood as between one laborer and another, and never more than perfunctorily frowned upon by way of public gesture. But the mores of a civilization comprise the decorously tailored side which each class presents perpetually to all the others.

On this side of the rich and strong we see what it were best for us to take to be their life and their behavior; and they are usually quite scrupulous about presenting this side to public view, resenting prying politicians passionately. Here we see their great astuteness in planning for the national good; the tremendous risks they take in investing their unspendable surplus in precariously youthful industries; the extraordinary refinement of their life among people of talent in the arts, actresses and the like; the tremendous contribution to the culture of the nation of their private ownership of all the fruits of genius. So also the middle classes, whose place it is to aspire but seldom to attain, present to the lower classes a similarly decorous appearance of industry and loyalty. As for the lower classes, whose place it is to labor perpetually without reward: it is well that no one should labor without hope. Therefore, we seem to say, let them look to us for a model of assiduous and unquestioning application and the self-restraint that leads to moderate and well-deserved success.

Knowing one's place is not class-consciousness. Nothing could be more dangerous or more unlikely in this scheme than class-consciousness. We often hear it said by misguided radicals that the rich are class-conscious, and that the poor can prevail against them only by becoming class-conscious too. It is not true. The rich are not class-conscious; they are socially conscious. Not that they are profoundly altruistic or transparently candid. When they are endeavoring to have the inheritance tax abolished, for instance, they may be frankly selfish in their secret hearts; and the bland arguments to the effect that such taxes should be state and not federal assessments may be wholly specious, a mere prelude to tax-dodging. But selfishness and speciousness are not confined to the rich; and it still remains true that the genuine convictions of the rich are shared by the entire nation. They believe that the inheritance tax is "socialistic," as it is; and they believe that a nation can be healthy only when the rich are growing richer all the time. Every one believes this. It is the gospel of the middle classes and the fundamental tenet of all successful trade-unionism: the sacred right of private property and capital accumulation. The rich are not conscious of antagonism toward other classes, certainly not of any least opposition to the established order.

The class-consciousness which radical agitators urge upon the working classes (however unsuccessfully) is a very different thing from the selfishness which is ingrained in the whole system of social emulation. These agitators would have the upper class abolished—wiped out. The rich have no such notion. They are not ambitious to have the poor wiped out. The gospel of the proletariat, moreover, calls for the nullification of the whole system of private property, class stratification and social emulation. It is not really the antagonism of one class to other classes; it is the antagonism of utopian fanatics to the entire existing scheme of mores.

Hence the horror in which the "class-conscious" radical is held. He is not denounced because he represents the poor. Obviously he does not represent the poor in fact; and moreover the rich love the poor. They are continually doing nice things for them. The understanding between the rich and the poor is quite complete: the rich are glad they are rich and the poor are only sorry they are not rich instead. This situation may involve a certain amount of individual friction; but the concord of social classes is extraordinary. The same mores bind them all: they all prefer the higher pleasures. The appalling thing is that radicals do not. They scorn the rich and all their ways; and when they come into power, as they have done in Russia, instead of donning purple and fine linen, they wear rough home-spuns, work twenty hours a day, and live on bread and water. Mr. Chicherin, when he comes to Geneva, wears a frock coat and a silk topper; but he manages to do it with such an ironic flourish as to make all the rest self-conscious in their finery. This is no way to act, and is naturally intolerable to the righteous, rich and poor alike. The rich are not the enemies of the poor; they are the ideal of the poor. **The radical is not the enemy of the rich; he is the enemy of society.**

CONTROL BY EMULATION

We sometimes hear it said that the poor have nothing to lose but their chains. But what are the chains that bind them? Not the imperial edicts of the plutocrats. The poor outnumber the rich ten thousand to one; and if they were bound by force alone, they would rise and crush the oppressor in the twinkling of an eye. What binds them is their own motives, their own interests and desires. They are men, however discontented—not mistreated cows. As men they look unto the high places whence cometh their hope. As men they scorn mere animal comfort. A discontented man is better than any cow because however miserable he dreams of better things: not the sterile ratiocination of desiccated philosophers; not the efficient creature-comforts—the glorified cow-barn—of the radical's utopia; but the real and concrete pleasures, the higher satisfactions which he sees actually before him, almost within his reach.

Thus it turns out that pleasure is real enough and that it is after all the ruling motive even in human herd behavior. But human beings, unlike animals, aspire always for a little greater happiness. What their happiness might be civilization teaches them and all the while reveals it to them, concretely, just a little way ahead. Civilization is a device for dangling before men's eyes the carrot of anticipation, and righteousness is the harness by which every zeal, however forlorn, is made to work.

It is only the radicals, the saints and the philosophers, who turn up their noses at the higher bliss. In a very real sense they are the enemies of society—they, with their eternal question: what doth it profit a man if he gain the whole world and lose his own soul?

CHAPTER XIII

THE RIFT IN RIGHTEOUSNESS

ALTHOUGH suffering and woe are the common lot of man, most men go through life without ever experiencing the pangs of revolution. Their socialization, or domestication, or enslavement—call it what you will—is too complete for them ever to doubt that they alone are the unchallenged arbiters of all their destinies. This is to the good. Civilization would be impossible if revolution were the general rule; and for most of us life would be insupportable if we were very fully aware of our condition as domestic animals. Whenever we hear a man cry out, "I am the master of my fate," we may be pretty sure that his head is bloody, though unbowed. No man becomes the captain of his soul except by mutiny, and even then the mutineer is apt to find all ports closed against him. Mutiny is a difficult, uncomfortable and dangerous expedient. No man can go through with it who is not first of all unusually sensitive to the indignities of his domestic posture, and then endowed with a rare capacity for bearing pain. He must be what sporting men call a "glutton for punishment."

But more important even than these personal qualities is the provocation. Mutinies do not arise on easy trips under humane and forbearing officers. Neither do revolutions occur in peaceful and well-adjusted communities. Yet it seems reasonable to suppose that sensitive and stalwart souls are no rarer in one place than in another. The raw materials of moral heroism must be commoner than the small number of accomplished moral heroes would lead us to suppose. On mature reflection we are therefore driven to recognize the fact that the effective cause of revolution is an intolerable situation. The man who cries, "I am the master of my fate! I am the captain of my soul!" really means to tell us, in this quaintly emphatic way, that he is in the grip of overpowering forces. He may be, in a certain sense, the captain; he is standing at the wheel; but he has been struck by the gale of circumstance, and he is flying before the wind at a terrific rate. The revolution consists simply in this: that his direction is now opposite to the one in which he originally put out from port.

There is a positive intellectual advantage in recognizing that all individualistic behavior, including moral individualism, follows the lines of social cleavage. One can then understand how it has come about. It has come about as part of an intelligible process. The case resembles that of the volcano. Formerly it was supposed that volcanoes were a primary occurrence, an explosion from within the earth, by which tensions were released and adjusted; and that earthquakes and the more gradual settling of the surrounding country were the effect of volcanic action. This may be compared to the conventional theory of moral leadership. The moral leader provides the explosion; he shakes the surrounding country; and the configuration of the continents is molded by his personality. This is a very nice theory, if we assume the volcano; but it leaves the causes of volcanic action, if causes we must have, decidedly hazy and hypothetical. Recent geologists, therefore, have not been content to receive their volcanoes as a revelation. They have proposed that the volcano is a secondary or even tertiary phenomenon. What is occurring in the first place is rising and settling upon a continental scale, along certain well-defined lines of cleavage. In the course of these major adjustments, minor and local disturbances sometimes occur, due to minor and local conditions. The character of the strata in some locality—the presence of volatile materials and the like at the point of cleavage—bring about a sudden local slip, the sudden slight emergence or subsidence of certain small features of the continental landscape, a slight tremor of the surrounding country, and occasionally a volcanic explosion as the result perhaps of the sudden passage of water through some newly opened fissure to a region where great heat has been generated.

This metaphor serves as well as metaphor can to illuminate what the geology of civilization is. The important adjustments in civilization are the adjustments between the continents; and whereas those adjustments of the earth are ruled by certain large but simple laws, ruling the comparative weights and densities of continents and oceans, the nature of crystalline structure, the geometric character of the sphere, and the like, in precisely the same fashion are the continental divides of civilization indicated. The passage from feudalism to industrialism is such a continental divide. We do not know as yet the extent or even the character of the continent which, in industrial society, is emerging slowly from the ocean of the past. But we do know that it is a new Atlantis, a different world from the one we have inhabited before. It has probably been brought about through the operation of certain large but simple principles: the diffusion, penetration and cumulative upward push of mechanical techniques; the interplay of the relative weights and densities of mechanical techniques on one side and of customs and superstitions on the other; and plain political geometry, such as the relation of the angle of literacy to the curve of the electorate.

The emergence of a continent or a culture affects an entire people; but it affects most violently those who live along a continental divide or those whose occupations cross an area of social disturbance. Because they live a nomadic life of great emotional stimulation, stage people and other artists are peculiarly susceptible to the forces which are disturbing our emotional and especially our sexual affairs. Some one has remarked that the much-discussed companionate marriage has been in actual practise for years; but as every one knows, it has been practised chiefly among these professions. Scientists and other students are peculiarly exposed to the hazard of religious doubt. Economists and sociologists are liable to become skeptics on the subject of the tariff and the gold standard; children, about the sanctity of the home; and working men, about the holiness of captains of industry.

This is not to deny the importance of the personal equation. Not all people who live on the rim of a volcano behave just alike; some move when an eruption threatens, and some stay from an excess either of inertia or of daring. The feeble-minded can often live through any crisis without being much affected, somewhat as Timothy Forsyte lived through the European War without knowing about it; and "hyperthyroids" become excited on slight provocation. But even though we assume that excitability is a pathological condition, due to glandular unbalance, still the activity so promoted may receive its direction wholly from the social environment. The man who in one age becomes a Mohammed or a Gautama will in another situation be a Napoleon or a Lenin. Peculiar persons flourish at all times; but different types of peculiars thrive in different situations.

In all situations, however, their lot is much the same. During the Augustan Age, Jewish civilization was approaching the zenith of instability, an instability which was marked by various signs and portents but most especially by the spectacular disturbance produced by a young renegade carpenter who had laid down his tools and become a wandering agitator. His experiences are most enlightening. At one time, for instance, he and his little coterie were passing along the countryside on the Sabbath day. They were without food; the disciples plucked the grain that was standing in the fields, rubbed out the kernels in their hands and ate it. But among this people there was a religious law against doing any useful labor on the Sabbath day. Consequently such of the righteous as were about, instantly spying and resenting this infraction of their laws, as righteous people always do, made indignant representations to the Prophet.

The Prophet was a peaceable man, and he wished to avoid trouble; so he gave these righteous men a soft answer such as occasionally turneth away wrath. He cited a precedent. The righteous are always strong on precedents. Therefore he pointed out that their greatest racial hero had once, illegally, eaten the sacred bread of the temple when he was hungry. He also tried to appeal to their sense of the practical. Righteous people are always intensely practical. He pointed out that the Sabbath was made for man, not man for the Sabbath.

But this, I fear, was a tactical blunder. In the first place it obviously implies that good sense is more important than formalities. That is not likely to please people who dote on formality. In the second place and only less obviously than that, it clearly implies that the Sabbath—since it exists for the convenience of man—is a man-made institution and therefore subject to amendment, a thought which is sure to terrify all staunch supporters of the status quo. People who want nothing changed will have it that things have always been the way they are and people who insist that things have always been that way refuse to admit the possibility of change. Nothing could enrage them more than the Prophet's calm assumption that even the holy Sabbath is subject to amendment.

At all events these pillars of righteousness made a special point of following the Prophet up to see what other, actionable crimes he might perpetrate. They had not long to wait. Very shortly a man with a withered hand put in an appearance and requested to be healed. This at once presented a fine point of law. One might argue that it is necessary to eat. But how could one justify breaking the Sabbath to heal a man who must have been a chronic sufferer for years, a man for whom a day's delay could make no conceivable difference? The point was too good to be missed, and these righteous men raised it at once.

In answering them the Prophet once again tried an appeal to their own peculiar mentality—an appeal, this time, to their cupidity. "What man shall there be among you," he said, "that shall have one sheep, and if it fall into a pit on the Sabbath day, will he not lay hold on it and lift it out?" The righteous always know how to make money on the Sabbath day. But of course they do it quietly. With all their knowledge of the law they know just how to have their servants do the actual work of extricating the sheep so that their own hands are technically pure, or how to date their checks ahead, or how to endow a church and thus get a special dispensation when they need it. But above all they know how to keep quiet about their little irregularities. They never make a public parade of their Sabbath breaking; and if anything should ever leak out about it, they would have their quiet explanation ready. The last thing they would ever think of doing would be to defend Sabbath breaking as a general practise. Consequently the effect of the Prophet's words was immediate and complete. "They were filled with madness, and communed with one another what they might do to Jesus." In the end they crucified Him. In confusing their peccadillos with His own blasphemy, He had proved Himself a moral pariah.

Generally speaking, people prefer murderers.

THE RIFT IN RIGHTEOUSNESS

Not the least interesting aspect of the legend of the crucifixion is the cry of the chief priests and the mob, "Away with this man, and release unto us Barabbas (who for . . . murder was cast into prison)." The curious thing, considering the vituperations we expend on the individualist, is the leniency with which sinners are regarded among us. The reason is that the sinner has broken the law but not flouted it; whereas the pariah has flouted the law, even though he has never broken it. In breaking the law the sinner even pays it a delicate compliment. He calls attention to its unchallenged sway. The sinner, moreover, bears evidence to the existence of temptation in the world, and therefore to the necessity for law. Sin is not a competitor of salvation: the antithesis of salvation is unbelief.

The Gresham's law of righteousness is: bad morals define good. Consider, for example, the case of prostitution. The righteous have a saying to the effect that the prostitute is the guardian of the home. Among many peoples her function is explicitly a sacred one; and even among us it has a quasi-sacrificial character. The prostitute is like the sacrificial goat of the Hebrews:

all the sins of the community are put upon her, and she is driven into the wilderness. In the drama of salvation she represents temptation. To have commerce with a prostitute is sinful—admittedly so. Hence the very existence of so indefensible a being serves to emphasize by contrast the sanctity of matrimony.

This is not all. The contrast between prostitution and extra-marital love is even more vivid. Prostitution is sinful; extra-marital love is positively wicked. Prostitution is conventionally secret in the sense that the community pretends it is not there. But extra-marital love is surreptitious, or if it isn't, so much the worse; it ought to be! The reason for the contrast is that prostitution, however squalid, is not a competitor with marriage. It can not be because it is squalid. Its squalor is its one redeeming trait among the righteous. Extra-marital love, on the contrary, is a direct blow at established morality; and the more idyllic its concomitant conditions, the greater is the force of the blow. A community must tolerate prostitution out of deference to original sin; and prostitution can be tolerated because sin is the road to salvation. Prostitution is the best of all arguments for holy matrimony. This was the view of no less a theologian than Saint Paul. To righteous eyes the most heinous aspect of "free love" is its scandalous pretension of innocence. It actually presumes to improve upon public morals. The abandoned lovers actually give out that they are following the guidance of their consciences. They have even the cheap cynicism to call attention to the canon of free will. They affect to believe in the importance of sincerity in moral behavior, and they have the effrontery to propose that they are right in giving way to their wicked impulses and that the community of righteous men are conventional, hypocritical and wrong. Hence they are not to be tolerated.

The issue of sincerity—which is our old friend "free will" in disguise—is a peculiarly troubled one because it is abstract. In the abstract every one believes in sincerity, just as every one believes in free will. But in the concrete we want people not merely to be sincere but to be sincerely like us; just as in the concrete we insist not so much on the freedom of men's choices as that their free choices shall be righteous ones. To be truly righteous one must conform freely and sincerely to one's own individual conscience; but in actual practise this means that a righteous man is one who does freely, sincerely and conscientiously just what his neighbors do. What other evidence is there that a man is acting conscientiously except that he does what all other conscientious people do? Obviously it is impossible to take his unsupported word in matters of community importance. If we did, what would become of morality? The ground-work of morality is an organized and stratified society. What would become of organized society if every man behaved according to his own individual inclinations? Everything depends upon our all acting in concert, conscientiously, from our own individual inclinations. The essence of free will, and sincerity, and individualism, as the righteous understand these things, is that every one shall behave like every one else of his own free will, and woe betide the man whose conscience is smitten with the curse of originality. The authentically righteous takes no risks. Prophet or not, he will have no place to lay his head.

Viewing sincere individualism in this light,

we see no inconsistency in our bringing pressure to bear on the youth of the land to assume their predestined rôle—as regular Republicans—as young as possible. Could we do otherwise? Can society foster strays and welcome revolutionaries? The difficulty raised by the case of the moral individualist is both real and acute. The common saying is that society can not have two sets of laws, one for ordinary people and another for individualists; and this is true. But it does not go to the bottom of the difficulty. It is not on account of their behavior that we dislike peculiar people but rather on account of their opinions. In fact, their behavior is usually negligible. When we are dealing with social revolutionists, for example, we almost invariably find no overt acts whatever for which we properly can hang them. To get justice done, therefore, we are obliged to engage *agents provocateurs* to incite our mild revolutionaries to some actionable crime, or else we must find some convenient crime that has already been committed and by a little dextrous manipulation of the evidence, "hang it on them." What real harm was there in a dozen disciples picking a little grain and eating it on the Sabbath day? None. Not even Pharisees could make a cause of war out of that. They were obliged to follow the situation up, to make use of the man with the withered hand in order to provoke Jesus into flagrant blasphemies. Ultimately it was even necessary to fabricate a revolution in order that Jesus might be executed as its instigator. This is a technique with which all departments of justice are familiar. The simultaneous release of an actual and convicted thief adds a culminating emphasis to the contrast between thoughts and deeds.

The fact is we do really have two sets of laws, one for ordinary people and another for moral pariahs. The moral pariah—we say; and thousands of good people said it recently about two Massachusetts anarchists—is not entitled to the ordinary safeguards of the law. What he has or has not done is immaterial. If a man respects the sanctity of the home, he may keep a dozen mistresses and be a pillar of society. Bernard Shaw once wrote of the poet Shelley that if Shelley's opinions had been different he might have had illegitimate children in every county in England and no one would have cared. The difficulty arose when Shelley flouted conventional hypocrisy, defended extra-marital love, and acted on his own convictions. This made him a monster of iniquity. Yet, as a professor of literature has written recently, of Shelley's actual sexual behavior "little more need be said than this: At the age of eighteen he eloped with Harriet Westbrook, who threatened suicide unless he rescued her from the tyranny of school; at the age of twenty-one he deserted Harriet whom he believed—perhaps mistakenly—unfaithful to him, and eloped with Mary Godwin, whom he loved; in the few other instances if any —for the evidence is inadequate—in which he yielded to physical passion, he was never, so far as is known, either cold, or cruel, or cynical." For a man endowed with wealth, an amiable temper and physical beauty to a remarkable degree, this record will bear comparison; and for innocence of selfishness and brutality, who of us, monogamous or not, can match it?

Frequently the moral pariah is a man who openly defends what most people secretly have done. Some years ago the Russian radical, Maxim Gorki, was expelled from America by the general refusal of American hotels to receive him in company with a woman to whom he frankly was not married. The hotels were of course supported by public sentiment and the law. The law forbids hotels to assign to the same rooms persons who have not registered as man and wife. This pious principle accounts for the prevalence upon hotel registers of "John Smith and wife" and "Tom Jones and wife." Maxim Gorki, however, refused to play the "butter-and-egg-man's" game and was obliged to return abroad to the haunts of open immorality. Whether, in his refusal to connive in the scruples of American hotel clerks, he was moved by an excess of innocence or an excess of cynicism, remains a question. Such a question dogs every moral individualist. The individualist's motives in avowing what others conceal are open to two interpretations. One is made by the public; the other by his biographer. According to the public, he is a monster of iniquity, so abandoned that he has at last forfeited all sense of decency and taken to a life of open crime. This was the Pharisees' view of Jesus. But according to more sympathetic interpreters, he is a man of unusual innocence, honesty and candor; a man for whom concealment is impossible; a man so free from guile that he scorns to save himself from public disapproval by obliquity of any kind; a man wholly without taint of common hypocrisy—in short, a super-moralist.

As to the treatment of the moral individualist it makes very little difference either now or later which of these views is the correct one, or whether either of them is. When public opinion is inflamed against him, a super-moralist is not likely to profit by an appearance even of extreme innocence and candor. The more guileless he appears, the more diabolical must be his character—so the public reasons; and posterity is equally undiscriminating. Is our own view of Jesus any less extreme, in its way, than the view taken by the Pharisees? The fact is, we adhere to a religion of which we conceive Him to have been the founder. Mohammed excites no such feelings in our breasts. Those worthy souls—our schoolbooks call them "Tories"—who staked their fortunes and their lives on their devoted attachment to their mother country, may some of them have been sincere and guileless souls endowed with an extreme degree of moral courage.

Or it may be that they were only innocent. Perhaps they were just simple people in whom British loyalty had become unusually deeply set, so that they were blind to the obvious tenor of events. Perhaps Shelley was an innocent—like the guileless fool of Wagnerian mythology. He had, possibly, lived in a world of perfervid emotion and idea until he had entirely dispensed with the common shrewdness which enables most men to live with their ears to the ground. It may be that he had no notion of the effect his expression of opinion had upon the common run of people. Perhaps the super-moralist is only a morally one-eyed man. On this account, very likely, he is a picturesque feature of the landscape, like an erupting volcano or any other rare and violent phenomenon. Therefore we look upon him with awe and admiration or with awe and horror as the case may be; and therefore we are liable both to underestimate his perplexities and to overestimate his influence. We are apt to admire him more than he deserves or fear him more than he deserves, exaggerat-

THE RIFT IN RIGHTEOUSNESS

ing either in one direction or the other the actual importance of his "leadership." These enthusiasms and phobias are the stuff of which popular "movements" are doubtless generated. But popular excitement is itself a transitory and ineffective thing. It is difficult to know with the unimpassioned certainty of history what a moral leader is, and what the lead is that a super-moralist may give. As often as not a moral leader, like a political leader, is a man who stands at the head of the procession in a high silk hat: a leader by virtue of his ostentation and our submissiveness. We shall not do very well if we identify as leaders merely conspicuous personages. But if we look not for the person but for the leadership, we hardly manage to do better. How does one detect leadership except by top hats and batons? Looking backward we may say that a leader is one who has left his name and imprint upon the scene of his activity. But let us see. The name which resounds most voluminously in the world to-day is that of Amerigo Vespucci. He is described in the dictionaries of biography as "an Italian navigator from whom the American continents derive their name." Whither did he lead? To the naming of a new world.

In what did the moral leadership of Socrates eventuate? He is described in the copy-books by his exhortation, "Know thyself." But it is not clear that we have actually come to know ourselves, or that any one ever knew anything that he could have learned in no other way than as a result of the Socratic exhortation. And Jesus? But are we to use the criterion of Vespucci? "A Jewish carpenter from whom the religion of the occident derives its name." Have we followed Him, and if so, in what direction? What do we conceive the lead of Jesus to have been? Mr. Robert Briffault has recently reminded us that Saint Athanasius, the father if not the actual writer of the Athanasian Creed which is received to this day in the Greek, Roman and English Catholic Churches, declared "that the appreciation of virginity and chastity, which had never before been regarded as meritorious, was the one supreme revelation and blessing brought into the world by Jesus Christ"; and in support of this declaration—surprising enough to us, although many other fathers of the church appear to have made similar declarations—Mr. Briffault piles up a heap of Latin citations which the curious may investigate. But let us be frank. No other issue has ever provoked more controversy than the question: What was the leadership of Jesus? Why? Because we have all assumed His name and each of us has gone on about his own concerns. Therefore each of us finds himself under the necessity of showing that his concerns are those of Jesus.

For the moral leadership of an inspired prophet to turn out just the opposite from his intentions is the most natural thing in the world. The super-moralist is picturesque—like the volcano—because he affords a brilliant contrast to the familiar scene. Because he is picturesque people are attracted by him; and because they have been attracted they create a stuffed figure in his likeness and set it up at the head of their procession. It is no secret that the great success of Christianity in the latter days of the Roman Empire was due to the fact that it appealed to the proletariat. It extended salvation to the common people. It castigated the powerful and the rich. Woe unto you that are rich! Woe unto you that are full! Woe unto you, scribes and Pharisees, hypocrites! Blessed be ye poor. But has Christianity always continued to be an unmitigated blessing to the poor? Or has it, for many centuries been the chief bulwark of the rich, the full, scribes and Pharisees? Karl Kautsky, a German Jewish socialist, pointed out that the softening of the revolutionary rigors of Christianity is perceptible even between the earlier and later versions of the Life: how Luke's vigorous and simple language becomes, in Matthew, "Blessed are the poor *in spirit*"; and he laid an ironic finger upon the difficulties of interpretation which this discreet mitigation has occasioned.

The prophets of the French Revolution let much blood for liberty, equality and fraternity; they confiscated the great estates and distributed the land among the citizens; and the net effect of all their work is a system of peasant landlordism which is to-day the chief bulwark of French capitalism against revolution. Mr. Kautsky has no monopoly of irony. The Russian revolution is plowing the same furrow. Its two most concrete measures have been the same: distribution of land to peasants and a "new" economic policy. In a generation or two, what more will the fruit of soviet leadership have been? We shall observe with interest.

These words may perhaps be taken as discouragement of moral pioneering. So much the better! Regarded from the point of view of the general public, the moral pioneer (like the volcano) is a nuisance. Regarded from the point of view of a personal career, moral pioneering is a loathsome disease, always unpleasant and usually fatal. Regarded from the point of view of history, it is the leading irony of fate. Let no one rush into it under the impression that he alone can unfurl the banner of progress. This is the one situation in which he can be very sure that if he holds back, some one else will rush hectically in. But, alas, this excellent advice is wasted! The super-moralist is of all men the least advisable. Like a jet of steam from the volcano, he is the master of his fate, he is the captain of his soul; to which let the violence of his language testify.

CHAPTER XIV

LAW VERSUS ORDER

THE outcome of every revolution is another dogma. This is one of those ironies of fate by which, more than by any merely human regulation, our pretensions and our zeal are held in check. Restless intelligence marks out perpetually new activities; but habit, no less assiduous, speedily assimilates them. Is this our misfortune? Only if righteousness is a misfortune, civilization a mistake and even mankind itself a zoölogical mischance.

Nevertheless there is a difference of temper between conservationists and revolutionaries. One seems to be facing forward while the other faces backward; and when we put it this way, facing forward does seem to be decidedly more admirable. "All of moral philosophy," says Warner Fite in one of the wisest and most eloquent of recent works, "is an attempt to say how man differs from the brutes. . . . Moral action, I say, is thoughtful action, and that is sufficient. . . . I continue to reject the distinction of the good man and the bad, as a distinction morally irrelevant; and the discrimination that I have in mind is between the presence of moral significance and the absence of it; which seems to me to mark the critical attitude." This is the Socratic view of righteousness. It is the noblest idea moral philosophy has produced; for it puts order, which is an attainment and an actuality, above law, which is a heritage and a fiction. Law is the order of the past; order is the law of the future. In one sense the difference between them is the difference between knowledge and wisdom; in another sense it is the difference between cowardice and courage.

The nobility of this conception of morality consists partly in its contrast to the dull fixity of law and partly in its courageous assumption of the active rôle. Morality, it seems to say, is the process of adjusting, not the conservation of adjustments. The difficulty is not that the contrast between conservation and adjustment is unreal—it is real enough; nor that changing is less noble than being—it is not. But since changing is impossible without being, can it be more noble? Since men can hardly pioneer unless other men keep the home fires burning, how can we say that it is more noble to pioneer?

How much a man admires active processes and whether he prefers a stable environment is partly a matter of temperament; and this is partly a matter of bodily physique. The physical organism, though it may be ruled by unchanging laws, is carried on by a balance of two kinds of process which we may call building up and tearing down, or anabolism and katabolism. These expressions, for all their Greek derivation, are not very exact. They are of the nature of metaphors, drawn perhaps from the mythology of gods and devils, warring principles of light and darkness. Still, it is true that eating and breathing bring vital elements into the system, while work and trouble tear them down; and it is also true that some physiques assimilate better than others, while others, afflicted perhaps with some glandular hypertrophy or decay, wear themselves out abnormally. We speak of the dyspeptic temperament, not without justice, and even nowadays of the hyperthyroid temperament.

But even so, the classification of conservatives and radicals by their physiques can be carried to extreme. Some men, because they are dyspeptics, are hideously conservative, regarding every proposal of social reconstruction, however mild, as a personal threat against their individual digestions. Other men, with the same digestive difficulty, look upon every existing situation with gloomy eyes and can not be satisfied with anything short of a complete sweep. If a zest for intense activity is responsible for some reckless agitators, it is also responsible for some dictators and captains of industry. Illustrations of each type are obvious and innumerable.

What chiefly determines a man's attitude toward alterations in the mores is the point at which the social shoe abrades his foot. Unhappy couples are more tolerant of divorce than happy ones. Young people are more tolerant of contraception than old people and professional celibates. Women are more sensitive to feminism than men. But most especially, the rich, the powerful, the successful and even the merely hopeful are vastly more concerned to maintain the status quo than the poor, the helpless and those without hope as things stand at present. We sometimes hear the complaint against socialists and communists that they are ill-considered men, gauche, greasy and guttural. But what do we expect to find social revolutionaries, suave, cultured, delicately scented men, with an Oxford accent and ultra-ritzy manners? The poor, too, enjoy their caricatures. They see the conservative always as a paunchy man, grossly over-fed, over-decorated, under-exercised.

Such caricatures, however they are exaggerated, have always an element of truth within them. The class struggle is sometimes described —often contemptuously—as the opposition of the "haves" and the "have nots." But that is precisely what it is, for all the contempt with which we may wave aside the vulgarity of mere dissatisfaction of the belly. Dissatisfaction of the belly is a very real affliction, so real that those who feel it might with some justice claim nature on their side. Theirs is a struggle for existence, and if the rich are so swollen with their pride of place that they will oppose to the poor man's struggle to exist their determination to retain their diamonds and their opera at any cost to their less fortunate contemporaries, then—say the spokesmen of the poor—let it be a struggle for existence, a struggle to the death, the annihilation of all the iniquitously surfeited.

But these are states of mind, not arguments. Moral injustice and social inequality, though they may be the primary source of every revolutionary movement, can not by any trick of logic justify a revolution. A situation may be very bad; it may be so bad that those on whom it bears the hardest will come to feel that any change must be to the good; but this does not mean that those who feel so will be right, and that any change will be actually to the good. Changes have a way of surprising those who anticipate them, whether with delight or horror. The same thing is true of conservation. The perpetuation of a state of affairs, tolerable to some and intolerable to others, is not in itself a good thing by any but the orthodox, conventional standards of that particular situation. Not the feelings either of the happy or of the miserable make any condition good or the reverse, but only mores. What makes one social order righteous is the righteousness of that social order; and what makes a prospective change seem right is the mores of social revolution—the convention of the revolutionists, Jacobin, Marxian, or Soviet—with which the changes are anticipated. When the changes have arrived they are sure to be decidedly different from what was anticipated; but they will then be good and right for rich and poor; and it will then appear that the contented obstructionists of the earlier day were wrong judged even by the standards of the reactionaries of that ensuing order, just as our own rich and powerful deplore the crude obstructionism typified by Charles I, Louis XVI and even, perhaps, Nicholas II.

Therefore, shall we say, blessed are the peacemakers? It would seem so; at all events it does seem so to timid, nervously sensitive temperaments, and to those who live in "no man's land," the Belgiums and the Luxemburgs of life, that is, clergymen, professors, editors and such like "responsible thinkers." Such people are so placed that they feel with the poor and depend on the rich—hence their eagerness for compromise. Taking a cue from their predicament, they propose that since neither change nor status is a good thing in itself, each being relative to some dogmatic preconception, why should not both be tempered with mercy and informed with justice? In part this means nothing, but is just the kind of smoke-screen men must send up who can not say what they think and will not think what for the most part they must say. It is only the sort of eagerness for change which would have everything remain just as it is. But in part it expresses a genuine, though erroneous, idea—the idea that slow changes are better than fast ones. But this is only another absolute with no more to recommend it than the dogma that all change is good or the equivalent dogma that no change is good. If any change is allowable, why not a rapid one? Is destruction more destructive that accomplishes its whole effect in one blow, or is it preferable to cut off the dog's tail one joint at a time?

We have a great way in these scientific days of talking about scientific study and experimental change. Smoke-screen words, these, meaning nothing but dignifying caution with a university degree, giving an honorary title to pusillanimity. This way lies abstraction, the densest smoke-screen of them all. To make abstractions of the active processes of righteousness has many times been proved the easiest of intellectual feats. The materials are always ready. To use their language, such abstractions are the description of the process of alternating change and status. Such a process is perhaps the law of nature for mankind. We have spoken of the inorganic universe as ruled by stable principles; but in every fragment of the inorganic world adjustments are continually being made. Following the dualism of the body we might say that anabolism and katabolism are universal. A "stable" atom or molecule is after all a temporary and more or less delicate equilibrium of physical forces. Alter the equilibrium by applying a little heat and a disequilibrium occurs, an explosion, readjustment. What goes on in the natural history of civilization is nothing more than this: a continual passing from equilibrium to disequilibrium and back again. Perhaps it is but the social phase of a universal process, the metabolism of the cosmos.

Suppose it is. Suppose the procession of the mores takes place through the working of some universal cosmic law. What then? Is status preferable or change? Will the universal process perhaps be better served by those who pretend that nothing is "really" changing, since a universal law is being served? Alas, universal laws are neither good nor bad; they merely are; and men do not "choose" to run with universal processes, or not to run. There is a wide-spread illusion—one typical puff of the smoke-screenists—to the effect that we can "further evolution," or we can "oppose" it. The American Museum of Natural History should mount a reconstruction of a dinosaur "opposing" evolution! What is a law of nature, if one may oppose it? Surely the essence of the case is that every man plays his part in an inevitable process. If he is pusillanimous, he may in one case strengthen reactionism, or in another succor revolution; but he will always do one or the other. Not even by abstraction can we arrange for changeless change.

Erecting intellectual abstractions out of the cosmic processes is no good for the rough and tumble of righteous living, nor is making an ideal of the physical concept of equilibrium any better. Righteousness is always concerned not with cosmic processes or physical concepts but with concrete mores as they come and go. Their coming and going is a natural process: that makes no difference to the righteous. Equilibrium is a physical fact, too. We can even go so far as to say that morality is a struggle on the part of every individual to achieve and maintain his behavior equilibrium in a changing world. The individual is spinning in a vortex of external forces. The mores whirl him about in one direction; his physical circumstances deflect this eddy in another direction. The mores direct him to turn the other cheek; but his starving family and the local union organizer exhort him to energetic resistance. His problem is to keep his head above water in the midst of this gyration, to preserve his balance or individual integrity.

But what is the standard of individual integrity and the measure of balance—living a rich and fruitful life? Few men do that. In these days of psycho-analysis we are liable to give a wholly negative interpretation to this problem: to say that a man has maintained his equilibrium if he has avoided "disorganization," that is, if he has not gone crazy or partially crazy. But psychopathologists have never given us any standard of sanity except the normal, or average, or customary behavior of the ordinary righteous man who is "adjusted" to the mores. By what device, except the mores, do we convince ourselves that a "creature of impulse," even of conflicting impulses, is any better or worse, more admirable or less desirable, than a creature of habit? Which is better, the "introvert" or the "extrovert"?

Since the struggle for equilibrium in the midst of conflicting forces is a universal process, suppose we take an analogy from the inorganic world, say of chemistry. Unadulterated table salt is a chemical compound of sodium and chlorine—one atom of each combined to form a diatomic molecule of salt. The combination is exceedingly stable. As every one knows, table salt can be subjected to all sorts of treatment including heat and cold, solution and recrystallization, and contact with a considerable number of other substances, without losing its identity as sodium chloride. But the two elements, sodium and chlorine, are both exceedingly active. Common salt appears almost universally in nature, the oceans containing enormous amounts of it; but neither sodium nor chlorine is ever found except in combination with another element. As a pure metal, sodium reacts explosively with water and burns energetically in air. The violence with which chlorine combines with other substances is the basis of its use in chemical warfare. In recent years chemists have formed a theory of why these things are so. A neutral sodium atom contains eleven electrons outside the positive nucleus, which is an excess of one over the number which (according to one of the most interesting of modern chemical generalizations) can form a stable configuration. Chlorine contains seventeen, which is one too few to form another pattern which is stable. When they meet, therefore, chlorine seizes the odd electron from sodium, becoming a negative chloride ion, while the sodium is left with an excess of positive nuclear charges (protons). So the sodium and the chloride ions, by virtue of their opposite charges, cling firmly together—and all is well. Viewed in the light of this theory, salt might perhaps be considered "better" than its elements.

We might even imagine that salt is what sodium and chlorine are ambitious to become. We might possibly impute these sentiments to the cosmos and say that "nature abhors" free chlorine gas and metallic sodium, just as we used to say that nature abhors a vacuum.

But now we are talking alchemy, and that is just what we are doing when, and if, we say that "nature abhors" the introvert, the paranoiac, or the over-compensated inferior. Does she indeed? We, to be sure, abhor chlorine gas when we meet it on the fields of the Argonne; and we conventional people very morally abhor introverts and Œdipus complexes. But that is owing solely to our prejudices in each case. So far as nature is concerned, we are at liberty to think that nature abhors human flesh, since she has made it so unstable in the presence of chlorine, and that she prefers abnormal people (just as God preferred fishes to folks) since there are so very many of them.

To righteous people, well brought up, the temptation to be "constructive"—to bring the argument through to some positive and salubrious conclusion is very nearly irresistible. It would be very gratifying if we could pronounce, at the end of all this negative analysis, that—whatever the mores—good-natured people, free-spirited people, big-souled people, large-minded people, in short likable people, are after all the great ideal toward which all the mores strive. It would be so very gratifying to find a place to draw the line! So very pleasant to end on a theme beginning "None the less . . ."! But alas, it is impossible. Unless we are to end by closing our eyes to all that has gone before we are obliged to see that the kind of people we like are the kind of people who like us. As the tawdry doggerel reminds us, not one of us is so abnormal or so debased that he is not good-natured and big-souled under some circumstances; nor is any one of us free from harsh asperity under other circumstances. A good-natured person is one who enjoys what we enjoy, lives as we live, and in a word is like us. An ill-tempered man is a man who is not "one of us." The reason he is ill tempered is almost certainly that we make him feel that he is not one of us. And are we then to hold a mirror up to nature? Are we the ultimate standard? Our mores?

It would be some satisfaction if we could even pronounce with finality that change is better than stagnation, and come out boldly for revolution. But even that is impossible, under the premises. How unsatisfying it is to say only this, that change is no worse than status and no less inevitable, and compromise no better than either one! But that is all the facts allow us. In the perpetual movement of the mores from equilibrium to disequilibrium and back again we can not even say except by arbitrary designation which of a passing series is equilibrium and which is disequilibrium. Each is a rule unto itself; all are natural together and fully share the approval of the cosmos. Morality is law and order. By fixing order with the seal of its approval, law promotes disorder. By struggling free of that disorder, revolution promotes law. Such are the verities by whose grace men thrive and in whose name men pillory each other!

CHAPTER XV

NO MORE CRUSADES

THERE is nothing to be done about it. This is at the same time an annoyance and a comfort. As a very moral race we are given to crusades and welcome every opportunity to raise a banner with a strange device and march forth to conquer or to perish. But as an animal species we are naturally lazy and willing enough always to be let off trouble.

This has not been an argument against righteousness. Nothing that has been said could possibly weaken the force of mores in the world. Some particular mores may have slackened their grip and some others tightened their hold on some occasional reader. But any event may have that effect. Some reader may find himself by so much confirmed in the practise of reading books, or some one else may just here have reached the point at which his system will absorb no more; but almost any other volume might have had the same effects. An author can not be held accountable for these obscure chemical reactions. But where the whole argument has been constructed to show the all-pervasive, immanent force of righteousness, no one can fairly argue that its effect is to impair that force. He must first show that it is impairable; and to do that he must first upset the whole theory of righteousness that has been presented here.

But perhaps he will feel that it is his theory which has been upset by our disconcerting argument. What has been impaired is perhaps the force of his own personal beliefs. This occasionally happens. But the grief with which adolescents, of whatever age, lament their broken idols is properly known as sophomoric. Such feelings express the confused condition of the partially enlightened, like the tears of the child who has discovered for the first time that storks do not bring babies or that Santa Claus is only daddy in false whiskers. As enlightenment increases there comes a compensating realization that a generous father is a reality worth grasping and that the birth rate is not diminished by any impairment of the functions of the stork.

The first sensation of the youth who is losing his juvenile religion is the feeling that the reign of order in the cosmos and righteousness among men is calamitously coming to an end. But this does not always turn out to be the case. Strangely enough, the intellectual growing pains of adolescents seem to have little effect upon the world. Long after we have grown accustomed to the idea that there is no Santa Claus and the dull ache has ceased to throb, we sometimes are reminded with a start that children are hanging up their stockings exactly as they always did. After all, in a world where the average mental age is twelve, only a few ever pass through an intellectual adolescence and emerge into untroubled doubt; and for them there are certain compensations.

One is the sense of superiority which sophistication brings. With the benign condescension toward their juniors of older children who have outgrown Santa Claus, the sophisticated are able to indulge in a measure of contempt, more or less good natured. This is one of the avowed objects of education—to give the alumnus a feeling of superiority to the uneducated; and if skeptical enlightenment produces it, so far it is righteous and regular enough.

But the major compensation is the realization that nothing matters much. What so violently agitates the adolescent doubter is the awful question: "If my childhood god is not the One True God, what is going to become of the heathen?" Their position was bad enough when they were only unreached by salvation; but if there is no salvation, they have exchanged a mitigable purgatory for eternal damnation. Some little time is commonly required by persons in this state of mind before the realization dawns that if there is no cosmic efficacy in salvation by their One True God, it doesn't particularly matter whether the heathen are baptized or not. But when the realization does finally arrive, it is a considerable comfort.

One's contempt is then directed not so much at one's fellow men as at their preposterous beliefs and their ridiculously solemn rites, and most particularly at the parasites of civilization—reformers, crusaders, holy men, archbishops, grand dukes and presidents—who live by propagating those beliefs and so exploit the common man's devotion. We do not know if civilization is a mistake, since at least it has produced us; but we know that it is a parasitic growth.

The fate of the universe and the progress of mankind must be left to the shysters. After all, what matters most in human life is little things: meat and potatoes for dinner, the young lady across the way, and the condition of the crop; and fortunately we are guided in such matters by traditions so familiar that it makes little difference if they are only our traditions. The most important thing in life is the struggle to make two carrots grow where one has grown before, or to make a sentence bear some faint resemblance to what is in one's head. In fundamentals we are after all brought back very close to common man; and as this realization germinates within us, there grows also a humbling respect for the animal who can carry so many parasites and live.

THE END